Dirk Zielke

MEMS
Micro-Electro-Mechanical Systems

2nd Edition

Bielefeld 2021

2nd Edition 2021

Black and White Print

Author: Dirk Zielke

Further information, please see masthead

Copyright Dirk Zielke 2021

All rights, even the partial reprint, excerpt-wise or fully, storage in data processing systems and translation, are reserved. The use of brand names, trade names, trademarks, etc. in this work does not imply, even without special labelling, that such names in the meaning of trademark and brand protection legislation are free for general use.

ISBN: 9798746358158
Imprint: Independently published

Preface

Preface to the 1st edition

The field of Microsystems is a rapidly evolving topic. This is due to the increasing quantities of micro-sensors through their integration into smartphones and their manifold use in cars, as well as through the use of these sensors in new areas, such as medical technology. With the present textbook as a tool, the reader will be able to get to know the state of the art in this field, and to use successfully Microsystems in various applications.

The following textbook is based on the lecture module "Microsystems," which is held at University of Applied Sciences Bielefeld in the 6th semester of the bachelor course electrical engineering. The lecture module includes a practical course, which deals with the structure and the characterization of an acceleration sensor module. The instructions for this course are attached at the end of the book.

Review of the 1st edition by D. Dobkin 2018 published on Amazon
"This book contains a quick review of semiconductor processing targeted towards those methods used for MEMS devices. It's too brief to be very helpful if you aren't familiar with the field, but if you have some experience in silicon processing, then it is a very useful review of what MEMS people actually use. The second part of the book focuses on MEMS accelerometers, including some examples of layout, signal detection approaches, and optimization. A little discussion on pressure sensors is also included. Not much said about the rest of the MEMS world. So a good deal for a modest price."

Preface to the 2nd edition

During the continuous use of the book for teaching at the University of Applied Sciences Bielefeld, a revision became necessary after 5 years. This is reflected first of all in a large number of new sensors from the various manufacturers and thus with significantly improved parameters. But also in terms of technology and design a lot has happened in recent years.
In the 2nd edition of the book, therefore, all overviews were therefore updated and new trends added. In addition, the book has been expanded to include MEMS actuators and RF MEMS.

The title of the book has been changed to "MEMS" for the 2nd edition, as this better describes the content of the book for the English-speaking world.

Table of Contents

1 Introduction 7

2 Technology 11
  2.1 Silicon Wafer 12
  2.2 Semiconductor Technology 13
    Layer Deposition 15
    Photolithographic Structuring 19
    Etch Processes 23
    Layer Modification 24
    Technology Example: Diffused Resistors Produced by a Planar Processes 27
  2.3 MEMS Technologies 28
    Bulk Micromachining 28
    Surface Micromachining 33
    Wafer Level Packaging 35
  2.4 Packaging 37
    Wire Bonding 42
    Printed Circuit Board Technology 43
    Thick-Film Technology 48
    Soldering 49

3 Sensors 52
  3.1 Acceleration Sensors 53
    Structure of MEMS Accelerometer 55
    Static and Dynamic Behavior Model of a Spring-Mass System 57
    Capacitive Readout Principle 60
    Capacitance-to-Voltage Converter (CV-Converter) 62
    Parameters of Acceleration Sensors 64
    Application of Accelerometers as Tilt Sensor 71
    Application example: Condition Monitoring 75
  3.2 Gyroscopes 77
    Optical Gyroscopes 77
    MEMS Gyroscopes 78
    Parameters of MEMS gyroscopes 81
    Applications of Gyroscopes 84
    Inertial Navigation Systems (INS) 84
  3.3 Pressure Sensors 89
    Piezo-Resistive Evaluation 93
    Readout of Measuring Bridges 95
    Parameters of Pressure Sensors 97
    Applications of Pressure Sensors 99

4 Actuators 105
4.1 Types of Drives 106
Electrostatic Excitation 106
Piezoelectric Excitation 108
4.2 Optical MEMS 111
Micromirrors 111
4.3 Fluidic MEMS 115
Micropumps 115
Inkjet Printer 116
4.4 Acoustic MEMS 119

5 RF-MEMS 122
5.1 Oscillators 122
5.2 SAW-Filters 123
SAW Sensors 124
5.3 BAW-Filters 126

6 Practical Course 127
Experimental Setup 127
Acceleration Sensor ADXL202 128
Complete Circuit of the 2-Channel Sensor Board 129
Startup Procedure for the Sensor Board 132
Measurements of Static Parameters 133
Determination of Dynamic Parameters 135
Measurement of the Temperature Dependence 136
Inclination Measurement 138
Task 139

1 Introduction

MEMS (Micro-Electro-Mechanical Systems) penetrate more and more of our lives today, often unnoticed by the public. They are the key component in many applications, allowing electronic devices to contact the outside world and to interact with their environment. MEMS were originally born out of the idea to use the well-developed technology of microelectronics for mechanical systems. A not insignificant function is played by crystalline silicon, which, in addition to its semiconductor properties, has excellent mechanical properties.

MEMS are sensors and actuators, whose structural widths are in the μm range, and which are closely connected to an integrated electronic evaluation system. The electronics can be integrated on the silicon chip of the mechanical structure, or, as a hybrid variant, can be placed together with micromechanical components into an IC-package.

Resonant structures used for filtering purposes were one of the first applications. The thermally very stable vibration characteristics of mechanical structures were used to generate a vibration in an oscillator. This type of frequency-selective component has evolved continuously to the present day, and finds its applications in modern filtering and timing devices.

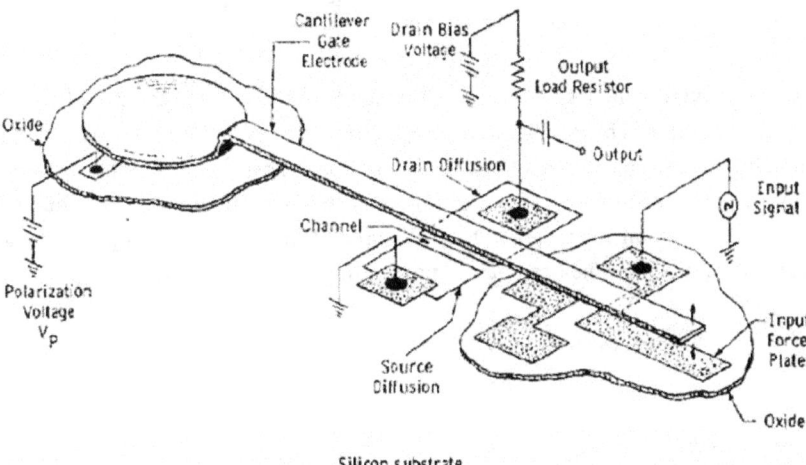

Fig. 1.1: Resonant field-effect-transistor[1] from 1967

In addition to some commercially interesting niche products such as the inkjet print head or the hard drive crash sensors, MEMS experienced their breakthrough with the widespread use of pressure and inertial sensors in the industrial and automotive sector in the 90s. Since then the number of sensors in cars has been continuously increased. Fig. 1.2 shows the current possible use of MEMS sensors in cars.

---

[1] The Resonant Gate Transistor, IEEE Trans. Electron Devices, March 1967, vol. 14, no. 3, pp 117-133

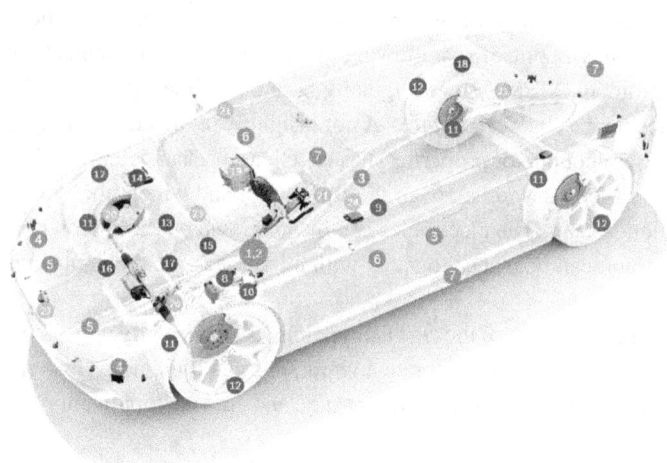

Fig. 1.2: MEMS sensors in passenger cars (Source: Bosch)

In the future development, self-driving vehicles will once again significantly increase the use of MEMS components in terms of quality and quantity. The number and complexity of the evaluation of sensor signals will increase significantly due to the autonomous mode of operation that this entails. If self-driving cars come to mind first in this context, a fair amount number of applications will be found in autonomous robots in industry and in household appliances. Vacuum robots are a good example of this. Due to the miniaturization of the sensors and the significant cost reduction, they can be used increasingly in households.

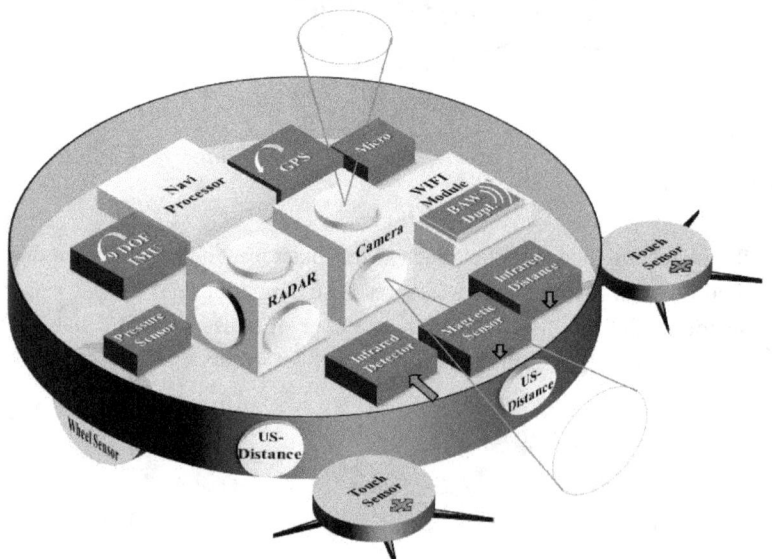

Fig. 1.3: Possible use of sensors in robot vacuum cleaners

Sales of MEMS received a further boost through their use in smartphones. In addition to the inertial and magnetic sensors for detecting movements, sensors for the measurement of environmental parameters (pressure, humidity and temperature) have also been integrated into them in recent years. In addition, at least two microphones and several RF-filter components can be found in many smartphones.

Introduction

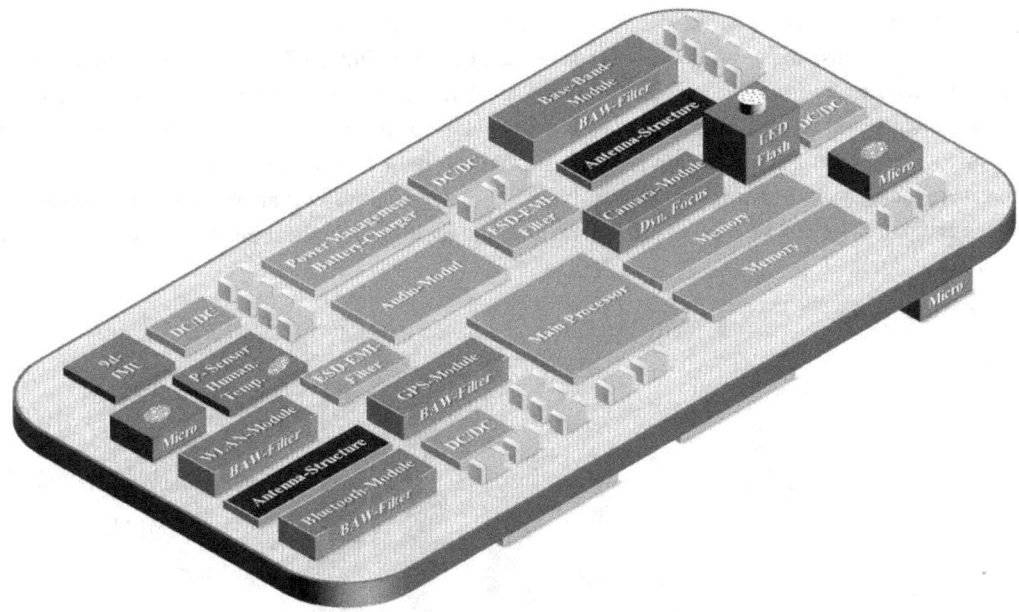

Fig. 1.4: Use of MEMS (shown in red) in today's smartphones

Since MEMS components are an integral part of the functionality of today's smartphones, the next stage of development will be in the field of autonomous sensors. They are characterized by a radio interface and their own energy supply. This means that self-sufficient sensors can be used inexpensively in large numbers due to their low installation and maintenance costs. An example of this is that of sensors for monitoring and supporting human functions. They can be integrated into a wristband or, in the future, into any piece of clothing.

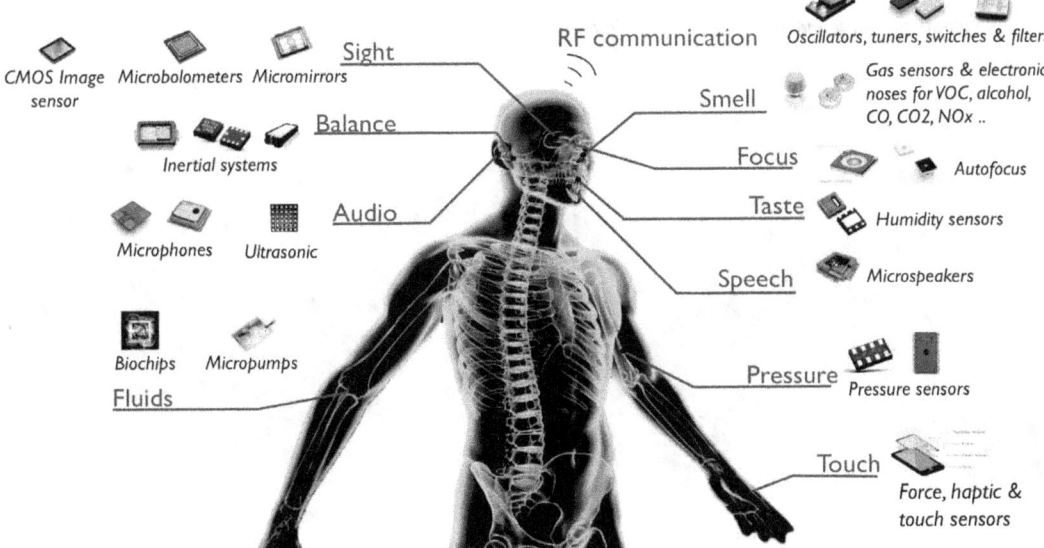

Fig. 1.5: Use of microsystems for monitoring and support of body functions [2]
(Source: Yole Developpement)

---

[2] Status for MEMS-Industry 2020 report, Yole Developpement, 2020

Fig. 1.6 shows the steady rise in the demand for MEMS devices worldwide. This development is remarkable mainly because the cost-per-sensor has reduced by approximately 95% since 2010. In addition, to the classic MEMS sensors (pressure, angular rate and acceleration sensors), other types of sensors, such as microphones, microbolometers, thermopiles, gas sensors and flow sensors capture an ever-widening market volume. Among the MEMS actuators, micromirrors in digital light processing applications (DLP) and the inkjet heads are the most successful applications. In addition, actuator systems are establishing themselves in microvalves and pumps. They are used mainly in medical and biological applications to provide doses of very small amounts of liquid.

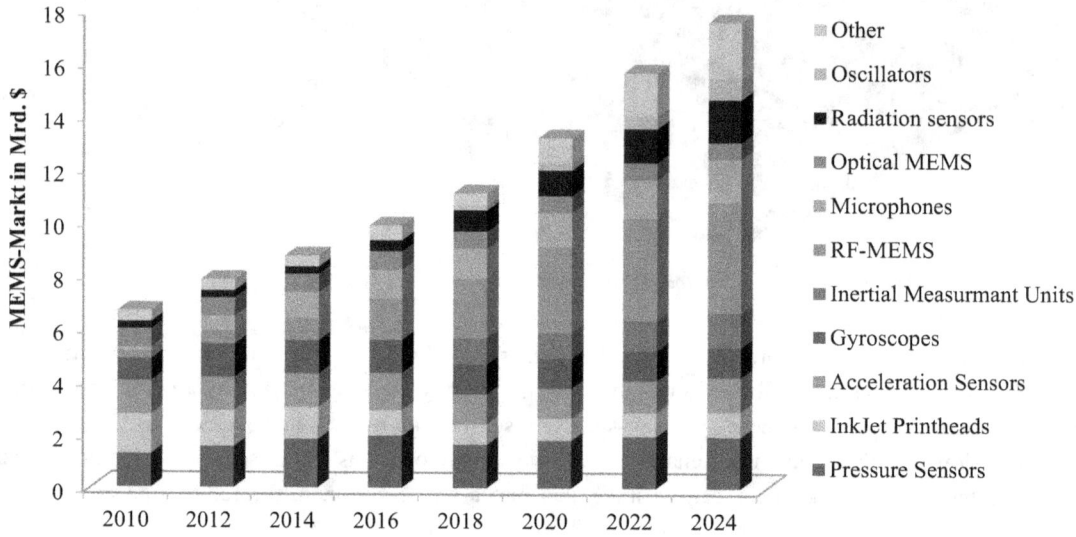

Fig. 1.6: Development of the market volume of the individual MEMS [3,4]

---

[3] MEMS-Market by Devices 2006-2017, 9/2013, IHS Technology
[4] Status des MEMS Industry Report, 2019, Yole Developpement

2 Technology

The manufacturing process of MEMS builds strongly on the process steps and technologies of microelectronics. At this scale, the manufacturers have access to a pool of technologies, which includes the production of silicon as well as the semiconductor technologies. The technologies are complemented by processes, which are necessary to create three-dimensional structure on MEMS devices. The situation is similarly for assembly and connection technologies. Here too, the standard processes of microelectronics can be used.

Fig. 2.1 shows the different height profiles of micromechanical and microelectronic structures. There are only very small steps in the surface in the microelectronic area. The step-height lies here below 1 µm. For micromechanical structures, the steps are significantly higher. In this example the profile height is 8 µm. For other microsystems the step heights can go up to the range of the wafer thickness (500-1000 µm).

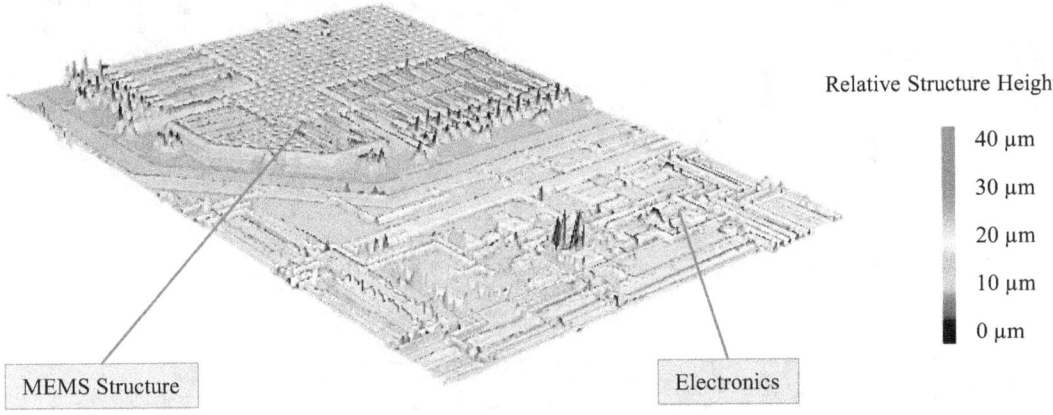

Fig. 2.1: Laser-scanning image of an acceleration sensor with integrated evaluation electronics

A recipe for the success of microelectronics is that the individual circuits are made in parallel on a wafer. Thus, even though the preparation of the silicon wafer is costly, the cost can be spread out across the large number of chips that are constructed on each wafer. The same concept is used for microsystem-technique. Here too, several microsystems are produced on a wafer, and were separated at the end of the production process.

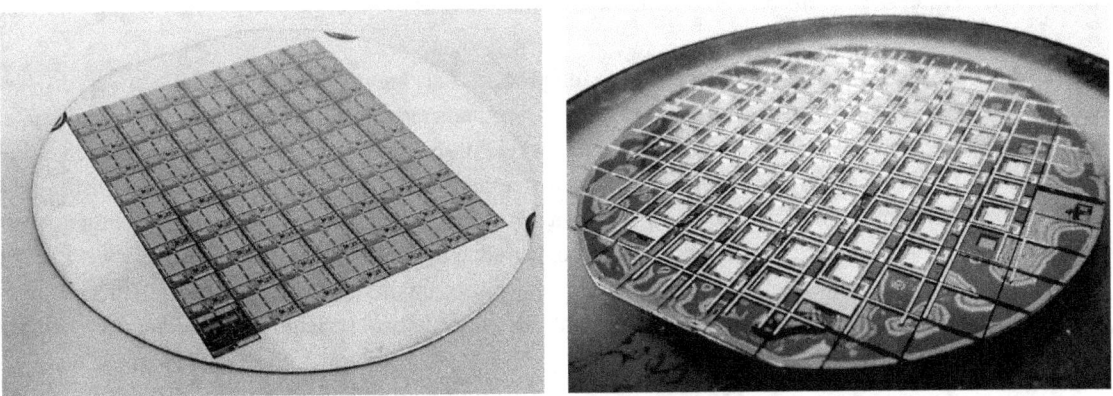

Fig. 2.2: Integrated circuits (left) and microsystems (right) produced as batch on a wafer

2.1 Silicon Wafer

The microelectronics used today one side polished wafers with {111}- and {100}-orientations and diameters from 8 to 12 inches. Current research activities are working on further enlarging the wafer diameter to 18 inches, to get even more space for the integration of circuits per wafer. However, in recent years, it has been shown that the cost reduction from increasing the wafer area cannot compensate for the higher investment costs as expected. For this reason, the introduction of the 18-inch standard for silicon wafers has not yet taken place despite enormous research expenditure (as of 2021).

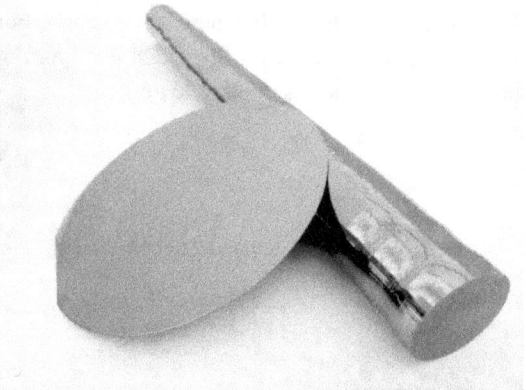

Fig. 2.3: 1-inch silicon ingot und 4-inch wafer

Fig. 2.4: 12-inch silicon wafers compared to a 20 cent coin

	4"	6"	8"	12"	18"
Diameter [mm]	100	150	200	300	450
Thickness [µm]	525	675	725	775	925
Flat /Notch	Flat	Notch	Notch	Notch	Notch
Bow [µm]	15	25	30	50	
Thickness variation [µm]	5	5	5	4	
Roughness R_A [µm]	0.01	0.01	0.0001	0.0001	

Tab. 2.1: Typical properties of silicon wafers with different wafer diameters

To adjust the wafer position during the production process, flats on one edge of the wafer are used. They point in a defined crystal direction. However, at large wafer diameters, this procedure is no longer workable. For a good usable flat, much silicon has to be ground away. For this reason, small notches are inserted in the material. Tab. 2.1 shows that wafers with a diameter greater than 4 inches have notches instead of flats.

As in classic microelectronics, silicon wafers are used for producing of microsystems. Currently 6- or 8-inch wafers are used. The much smaller dimensions compared to microelectronics (which usually use 12-inch wafers) are based on the lower number of pieces, as well as on the special demands of micro-system technologies. From today's perspective, the equipment for 3-dimensional structuring of mechanical elements is only economical up to a wafer diameter of 8 inches.

2.2 Semiconductor Technology

For the production of electronic components, a sequence of technological steps is used, in which four types of processes are constantly repeated:

- Layer deposition
- Photolithographic patterning
- Etching
- Layer modification

For most electronic components, only the first micrometers of the silicon substrate are used to build the electronic components (planar technology). Layers that are deposited on the substrate are mostly in the submicrometer range (10 nm - 2 μm).

To process the silicon wafers, it is necessary to ensure clean rooms with the lowest possible dust exposure, since every dust particle on the wafer leads to the failure of the respective chip. In modern chip factories, around 35 particles/m^3 (clean room class 10) are achieved. In order to achieve an even higher dust class during processing, systems are completely enclosed so that no particles can be brought in from the outside.

Fig. 2.5: Clean room production at Infineon (Source: Infineon Technologies AG)

Before looking at the individual microelectronics technologies in detail, the overall process will be demonstrated using a discrete diode, as shown in Fig. 2.6.

Fig. 2.6:   Chip photo of a discrete diode

Fig. 2.7 shows the process steps required to manufacture a diode. The overall process comprises 16 individual process steps, with the photolithography process only being shown in detail once.

An n-silicon wafer is used as substrate, on which a low-doped n-layer is deposited by epitaxy. This layer is monocrystalline, and contains all active areas of the electronic devices at the end of the process. As the next step, a layer of SiO_2 is oxidized onto the wafer, and is afterwards structured by photolithographic techniques. This layer acts as diffusion mask for the p-doping diffusion process, which creates the anode. As a final step, the sputtering and structuring of the aluminum layer, as well as the sputtering of the back contact (nickel/gold) is carried out.

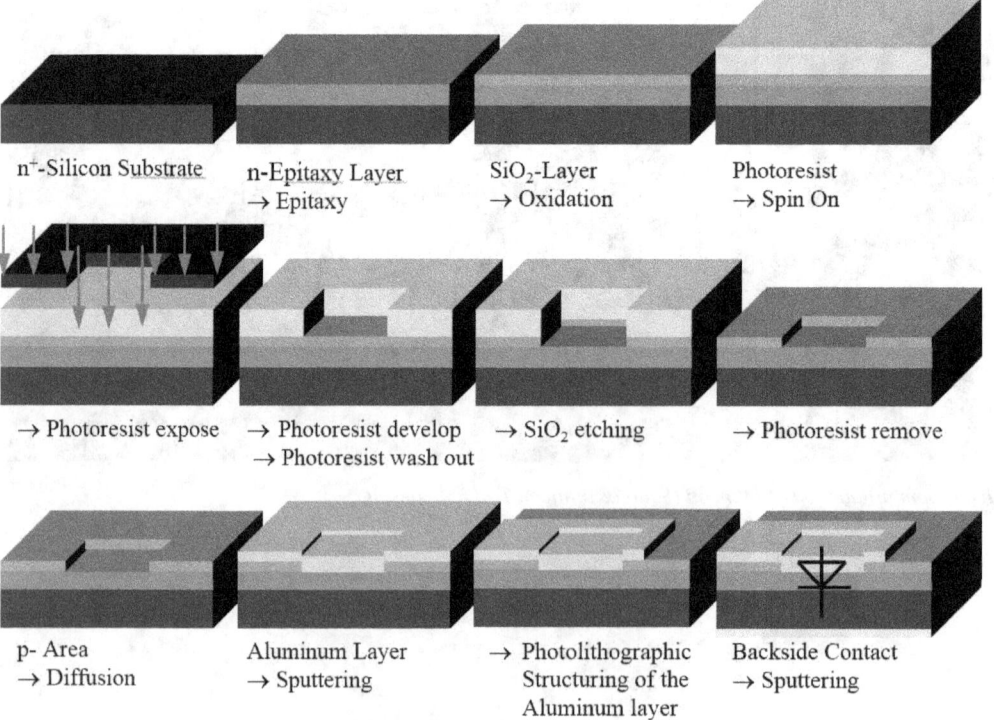

Fig. 2.7:   Process steps for the production of a diode

Layer Deposition

Layer deposition processes are in general applied non-selectively. This means that the entire surface of the silicon wafer is coated with a homogeneous layer. The layers have a thickness in the range of 30 nm to 2 µm. For this reason, this technology is called thin film technology.

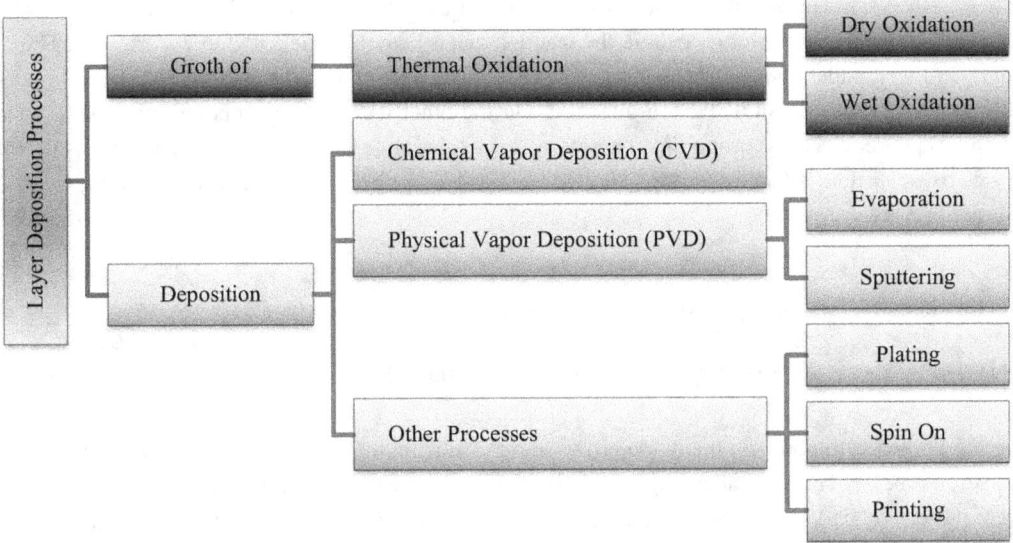

Fig. 2.8: Commonly used coating processes at a glance

Oxidation

The oxidation of silicon is one of the technology steps, which makes silicon technology so successful. SiO_2 is a very good electrical insulator, and is chemically resistant to most acids and bases. Furthermore, SiO_2 fits well to the lattice structure on the silicon crystal.

$$Si + O_2 \rightarrow SiO_2$$

Properties of SiO_2:
- Specific density of thermal SiO_2: $\rho = 2.2$ g/cm³
- Band gap: $W_G = 9$ eV, a very good insulator
- Breakthrough field strength: $V_D = 1000$ V/µm
- Good adapted Si-SiO_2-interface
- Good diffusion mask for most materials
- Very good etch selectivity between SiO_2 and Si

Fig. 2.9: Oxidized silicon wafer (dox = 170 nm)

Bare silicon surfaces have a natural oxide film of about 30 nm. This protects the silicon from a further oxidation. Only at higher temperatures (T = 800 to 1200 °C) is the mobility of the oxygen molecules in the SiO_2 high enough to diffuse through the already existing oxide layer to the silicon surface. At the beginning, a linear reaction-limited oxide growth is seen. However, at higher oxide thicknesses the growth of the oxide will be parabolic. Here the transport of the oxygen molecules to the reaction front is the limiting factor. That is why usually only oxide thicknesses below 1 µm are generated by the oxidation processes.

$$d_{ox} = -\frac{A}{2} + \sqrt{\left(\frac{A}{2}\right)^2 + B(t+t_0)} \quad [1]$$

with d_{ox} - Oxide layer thickness
A - Linear growth coefficient
B - Parabolic growth coefficient
t_0 - Describes the initial oxide thickness

There are two types of oxidation processes, dry and wet oxidation. For dry oxidation, pure oxygen is used as a reaction gas; whereas in wet oxidation, the reaction gas is heavily enriched with water vapor. The higher growth rate of wet oxidation is based on a less densely packed oxide layer, through which the H_2O-molecules can diffuse faster to the Si/SiO_2 reaction interface. On the other hand, the open-pored oxide layer of a wet oxidation has poorer electrical properties. Therefore, dry oxidation is used for thin oxide layers, as is required at the gate oxide of field-effect transistors; and the wet oxidation is used for thicker oxide layers, for example as a passivation layer.

Dry Oxidation: $\quad Si + O_2 \rightarrow SiO_2 \quad\quad B \cong 100 nm/\sqrt{h}$

Wet Oxidation: $\quad Si + H_2O \rightarrow SiO_2 + H_2 \quad B \cong 500 nm/\sqrt{h}$

for $T = 1100\ °C$

During the oxidation process the oxide grows into the silicon, through the consumption of silicon during the oxidation process. There are advantages and disadvantages in the process. An example is the "beak-effect" during the selective oxidation, seen in Fig. 2.10. Here the nitride mask is lifted off by the growth of oxide, and loses contact to the silicon substrate. On the other hand, the growth of the oxide layer into the silicon also can bring a benefit. Thus, the depth-level of integration in which the active components are built, is shifted into the silicon wafer. In this area are found far fewer imperfections (chemical impurities and crystal lattice dislocations) than on the original wafer surface, so much better device parameters can be achieved.

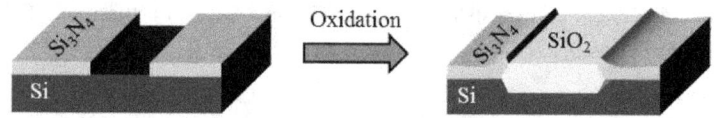

Fig. 2.10: Growth of the oxide during a selective oxidation process step

Oxidation systems consist of a heated quartz tube, a wafer carrier and the gas in- and outlet. For wet oxidation processes, the system is extended in addition by a bubbler, in which the oxygen flow is enriched with water vapor. To make the process economical, up to 500 wafers are oxidized at the same time.

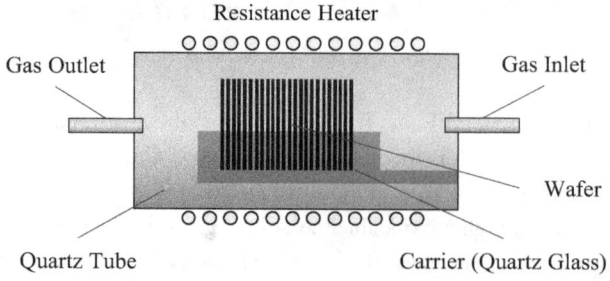

Fig. 2.11: Schematic representation of the construction of an thermal oxidation equipment

Technology

Chemical Vapor Deposition (CVD)

A universal method to deposit different types of layers is Chemical Vapor Deposition (CVD). It is based on a chemical reaction of a variety of reactants on the silicon surface. The reaction partners are fed into the reactor tube, and react on the hot silicon surfaces. The chemical reaction products form a layer during this process. Due to the different reaction partners, a wide range of temperatures and pressures are necessary to achieve an ideal layer formation. Ideally, the film-forming reaction should only take place on the silicon wafers. That is why the outer wall is often cooled. In this case, an induction heater is used, which heats only the silicon wafer, but not the wall of the tube or the wafer carrier.

A specific CVD process is Epitaxy. Different from normal CVD processes, where amorphous or polycrystalline layers are produced, Epitaxy is in the position to produce monocrystalline silicon layers. High processing temperatures and a monocrystalline wafer surface is required, because only under these conditions, do the deposited silicon atoms form a monocrystalline silicon layer.

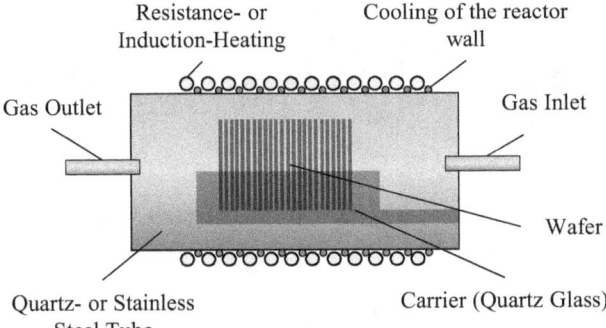

Fig. 2.12: Schematic representation from building blocks of a CVD system

Layer	Application Examples	Temperature [°C]	Process Gas
Si-Epitaxy	Active areas	1150 … 1280	$SiCl_4 + H_2$
Poly-Si	Conduction traces, Gate-electrode	60 … 650	SiH_4
a-Silicon	Anti-reflective layer	560	SiH_4
SiO_2	Dielectric	400 … 430	$SiH_4 + O_2$
Si_3N_4	Passivation	360	$SiH_4 + NH_3$ (plasma enhanced)
Al	Conduction traces	230 … 270	$Al(C_4H_9)_3$
W	Conduction traces	300 … 400	$WF_6 + H_2$
$MoSi_2$	Conduction traces	330	$MoCl_6 + TiCl_4 + SiH_4 + H_2$

Tab. 2.2: Example of CVD-processes

Physical-Vapor-Deposition (PVD)

The processes of evaporating and sputtering are grouped together under the term Physical Vapor Deposition (PVD). Both methods produce a very fine-crystalline structure, which is crystallized by an annealing process in most cases.

Evaporation

The evaporating technology represents a very simple process. The source material is melted in a crucible. Out of it, the material evaporates, and condensates afterwards at the wafer, where a homogeneous layer with a constant speed of growth is formed. Evaporative processes are very rare in today's technologies. They have been largely replaced by sputter processes.

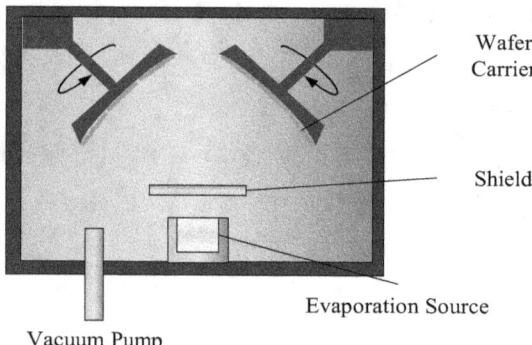

Fig. 2.13: Schematic representation of an evaporation system

Sputtering

The sputtering process is a very versatile application, as it is able to atomize a wide variety of materials. The only requirement is that the target material is not destroyed by the bombardment. Very often, the procedure is used for depositing thin layers of metal. Fig. 2.14 shows the structure of a sputtering coater. In a protective gas atmosphere (Argon, with a pressure range of 1 to 10 Pa) a plasma is ignited, which leads to the formation of Argon ions. These are accelerated in a high-voltage field, and hit the target material with high kinetic energy. Due to this collision, target atoms are released, and move to the wafer placed on the opposite side. There they form the required layer. Sputtering produces amorphous, relatively porous layers that can be densified through a subsequent annealing process.

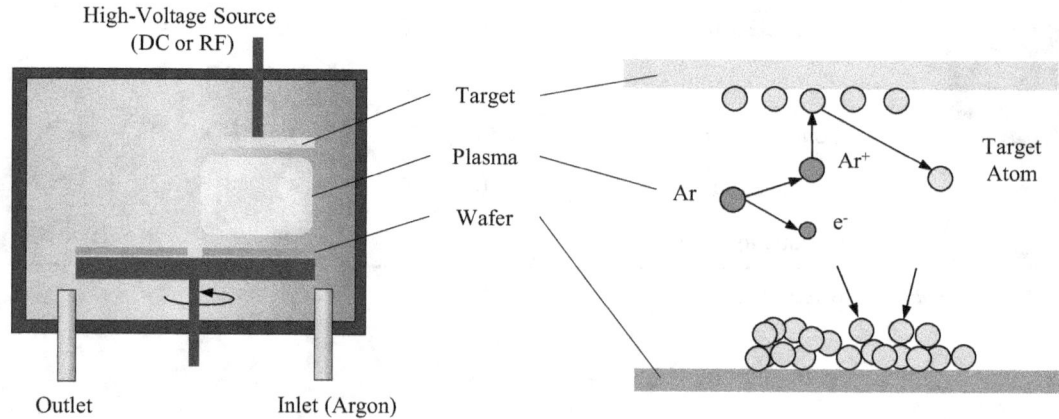

Fig. 2.14: Schematic representation of a sputtering system

Photolithographic Structuring

Structures of microelectronics and microsystems technology are produced by photolithography. In this process, a photosensitive coating is deposited on the silicon wafer and selectively exposed by means of a mask. Following exposure, the exposed areas can be washed away and thus a pattern is transferred from the mask to the silicon wafer. Fig. 2.15 shows the first photolithographic process of the diode's preparation, seen in

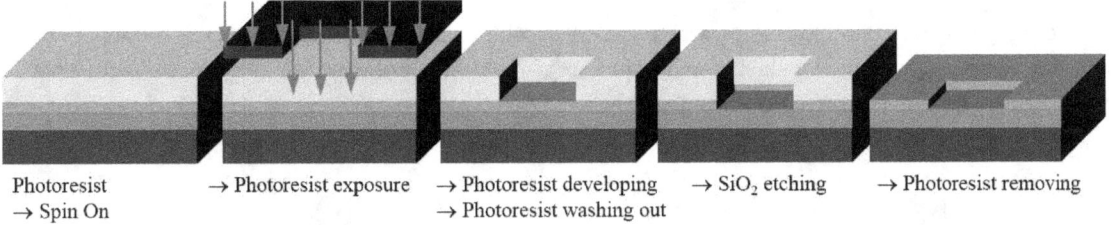

Photoresist → Photoresist exposure → Photoresist developing → SiO_2 etching → Photoresist removing
→ Spin On   → Photoresist washing out

Fig. 2.15: Photolithographic structuring of a SiO_2-layer

Very often, a silicon oxide or a silicon nitride layers are structured by means of a photoresist, and are used as the real mask for a following high-temperature process. In the example of the diode in Fig. 2.15, the structured SiO_2 is used as the diffusion mask, since the diffusion process takes place at 800 to 1100 °C. If the originally photoresist would be used here, they would be decomposed at this temperatures.

The exposure of the photosensitive resist causes a change in the structure of the polymer. As a result, the exposed areas of the resist can be washed out with an alkaline developer. On the other hand, the developer does not attack unexposed areas. At the end of the process, an image of the mask is created with the same polarity as in the photoresist (positive resist). For negative resists, the process is exactly opposite. Here, the exposure leads to a crosslinking of the resist and to a resistance of these areas to the developer. For this type of resist, the developer washes out the unexposed areas of the resist structure and an inverted image of the mask structure in the resist layer is obtained.

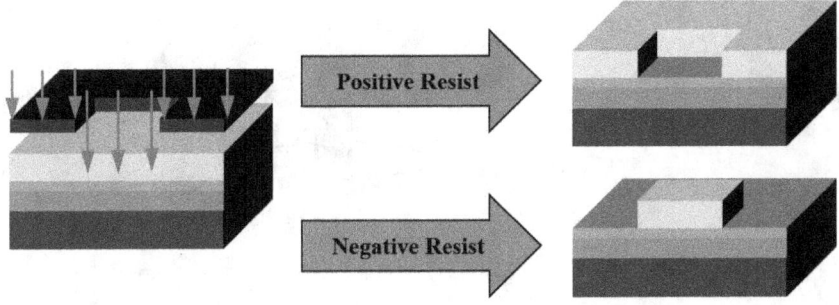

Fig. 2.16: Resulting images from positive and negative resist

After structuring of the respective layer, the photoresist is completely removed. This can be done simply with an acetone bath, if the resist is unstressed. If the resist subject to high temperature (T > 150 °C) during the exposing or following process steps, the photoresist is highly cross-linked and can only be removed by means of aggressive procedures (e.g., oxygen plasma etching).

Spin-On

Photoresists are spread over the whole surface of a wafer using the spin-on procedure. During this process, the photoresist solution is applied to the center of a rotating wafer and is spread outwards by centrifugal force. Excess resist will spin off at high speeds (2000 to 10000 rpm). Because the viscosity of the resist is increased by evaporation with time, the final thickness is independent of the spin time. Only the speed and initial resist viscosity are parameters for the resulting thickness. In microelectronics, photoresist layers from 0.5 to 2 µm are used.

For very structured wafer surfaces, which can occur with micromechanical elements, it is not possible to achieve homogeneous coating thicknesses, and especially good edge coverings, by means of a spin-on procedure. Here spray processes are used to apply the photoresist.

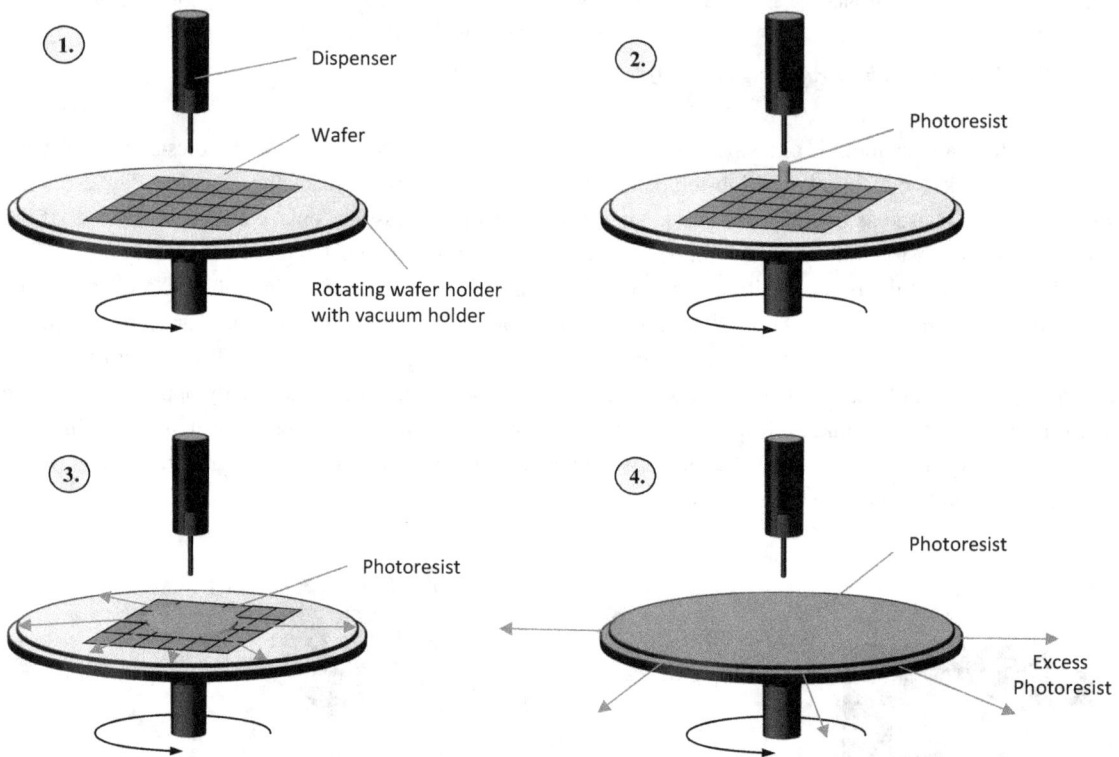

Fig. 2.17: Schematic representation of the spin-on process

Exposure Procedure

One of the simplest methods for a pattern transfer on the photoresist is contact exposure. During the procedure, the mask is directly placed with its chrome side on the photoresist; afterwards the photoresist will be exposed. Since there is no distance between photoresist and mask, refraction phenomena on the mask openings and the depth of field of the light source play only a minor role. Therefore, a relatively sharp-edged exposure of the photoresist is achieved with this procedure. The resolution limit is approximately 0.6 µm. A disadvantage of contact exposure is a relatively fast pollution of the mask, through its direct contact with the wafer. For this reason, contact exposure is no longer used on an industrial scale today. It has been replaced by the proximity exposure. With this method, a distance of 10 to 30 µm is set between wafer and mask. Thus, pollution of the mask through the contact with the photoresist can be excluded. Due to the distance between the mask and the silicon wafer, diffraction effects cause radiation to undercut the mask structures, and cause a higher minimum resolution with this process of about 2 µm.

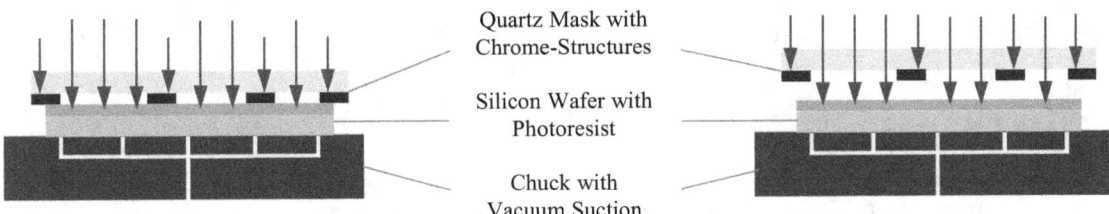

Fig. 2.18: Contact- und Proximity-Exposure

A very frequently used exposure procedure is the projection process. With it, the mask and the wafer are physically completely separated and only the figure of the mask is projected through a lens system on the wafer. By means of a reduction of the figure by an optical path, the mask does not create an image for the entire wafer. Usually a five times enlarged mask structure is used that contains only the information for one chip. Using a step-and-repeat mechanism (termed a wafer stepper), the whole wafer is exposed repeatedly with the chip image. The limit of resolution of such a projection process is approximately 40 nm.

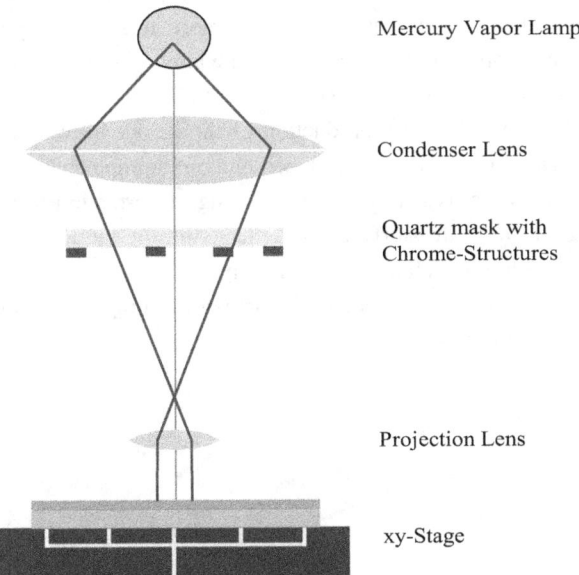

Fig. 2.19: Schematic representation of a projection exposure system

Wavelengths in the range of 13.5 nm to 436 nm are used for the exposure of silicon wafers. Therefore, the lower the wavelength, the higher the possible resolution. Fig. 2.20 shows the comparison between the minimum structure width and the wavelength used by the exposure systems in chip production. Usually, the smallest possible pattern width is adequate to half the wavelength of the exposure equipment. In 2005, the development of radiation sources below 193 nm stalled, meaning that exposure processes had to be developed that could circumvent this limit. One possibility is immersion lithography. Here, a transparent, highly refractive liquid replaces the air between the wafer and the last lens of the projection exposure system. In highly refractive media, the incident light's wavelength is reduced, thus making structure widths of up to 28 nm achievable.

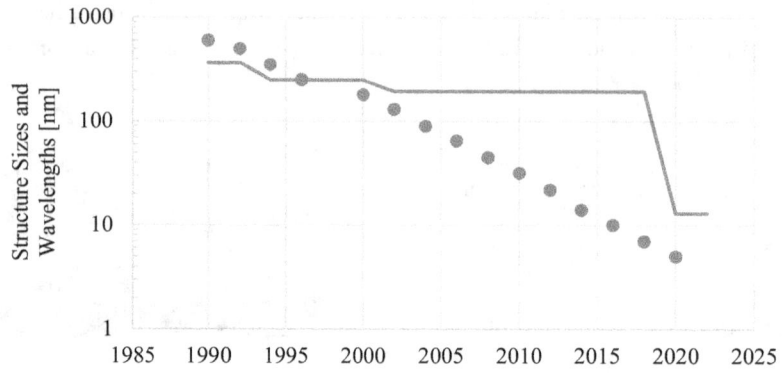

Fig. 2.20: Evolution of the wavelengths used in photolithography and the achieved resolutions

Another possibility to increase the resolution of projection exposure systems is the use of interference masks. Here, the interference pattern that occurs at narrow gaps is inversely included in the mask design so that the resulting interferences add up to a sharp-edged pattern in the photoresist. By 2019, these and other refinements of the exposure process led to a pattern width resolution of 10 nm.

From 2020, after a long development period, a light source with a wavelength of 13.5 nm became available (Extreme Ultraviolet Lithography, EUV). The radiation is generated by a sodium plasma and then projected onto the wafer by mirror systems. Due to the short wavelength, it is necessary to use a mirror projection system since quartz lenses are no longer transparent enough in this wavelength range. Furthermore, it is necessary to guide the entire beam path in a vacuum to exclude the absorption effects caused by the atmosphere. With the help of EUV lithography, resolutions of 5 nm are now achievable (as of 2021).

The described projection exposure is suitable for the effective reproduction of the layout information contained in the quartz mask on the silicon surface. The masks are also produced photolithographically. For this, a direct writing laser and electron beam lithographs are used to transfer the design information to the mask, in which an electron or laser beam is focused on the wafer surface and deflected in the xy direction. Electron beam exposure achieves resolutions of up to 25 nm, which means that minimum structure widths of 5 nm can be transferred to the substrate with an imaging ratio of 1:5. Due to the time-consuming procedure, direct exposure methods are only used in the production of photomasks.

Technology

Etch Processes

In order to structure the applied thin layers, various etching techniques are used. There are wet and dry etching processes. Dry etching processes are again divided into physical and chemical etching technologies.

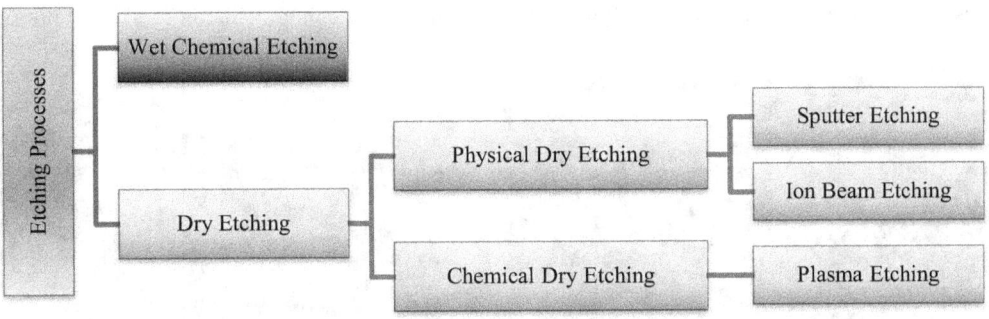

Fig. 2.21: Classification of etching processes used in microelectronics and microsystems technology

Etching processes are described quantitatively by three parameters: etching rate, anisotropy, and selectivity. The etching rate describes the amount of removal of material per time. If the etching process is ideally reaction limited, then a constant etching speed is seen and thus a linear ablation of the etching time occurs. This happens however only as long as the etchant is transported quickly enough to the etching surface, and the reaction products transported away from the etched surface. If this is not the case, the removal of the material could become non-linear with very small mask openings, and can come to a complete standstill in an extreme case. Chemical etchants are often high isotropic (AF ≈ 1) and selective (S > 100). Physical etching techniques, however, are usually strongly anisotropic (AF → 0), but have a poor selectivity.

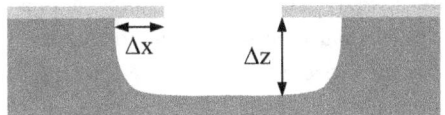

Ätzrate: $\quad r = \dfrac{\Delta z}{\Delta t}$ [2]

Anisotropy: $\quad A_f = 1 - \dfrac{\Delta x}{\Delta z} = 1 - \dfrac{r_H}{r_V}$ [3]

Selectivity: $\quad S = \dfrac{r_A}{r_B}$ [4]

with r - Etch rate [µm/min]
Δz - Vertical etch depth
Δx - Horizontal undercutting depth
A_f - Factor of anisotropy
r_h - Etch rate in horizontal direction
r_v - Etch rate in vertical direction
S - Selectivity
r_A - Etch rate of material A
r_B - Etch rate of material B

Material	Etchant	Etch Rate
SiO$_2$	HF / NH$_4$F	100 nm/min (25°C)
Polysilicon	HNO$_3$ / HF / CH$_3$COOH	15 µm/min (25°C)
Al	H$_3$PO$_4$ /HNO3	350 nm/min (40°C)

Tab. 2.3: Wet chemical etchants

Layer Modification

The annealing is the simplest form of layer modification. It is used very often to minimize layer stresses, for surface drying, or to support re-crystallization processes. Tempering processes take place usually in an inert atmosphere (N_2) to avoid unwanted oxidation processes on the wafer surface. Additional layer modification processes are diffusion and implantation. Both methods are used for doping of silicon semiconductors.

Fig. 2.22: Grain growth in $MoSi_2$-layers after 30 min at T = 620 °C, 800 °C, 1000 °C and 1100 °C

Diffusion

During the diffusion process dopant atoms (B, P, As) are brought into the silicon lattice to influence the conduction type of the semiconductor. The process is carried out at high temperatures (950 to 1200 °C). At these temperatures, the mobility of the dopant atoms is high enough to move into the silicon lattice. This is caused by its concentration difference. The dopants are made available over a source layer (diffusion out of an exhaustible source) or directly over the gas phase (diffusion out of an inexhaustible source). In both cases, it achieves a greater penetration depth by increasing the process time or the process temperature. For diffusion processes, the maximum dopant concentration is always found at the surface.

$$F = -D \cdot grad(N) \quad [5]$$

$$D = D_\infty \cdot e^{-\frac{W_A}{kT}} \quad [6]$$

with
F - Particle flux [mol/m^2s]
D - Diffusions coefficient [cm^2/s]
N - Dopant density [cm^3]
W_A - Activation energy of the diffusion process [eV]
D_∞ - Diffusions constant (depending on the dopant)
k - Boltzmann constant
T - Temperature [K]

Technology

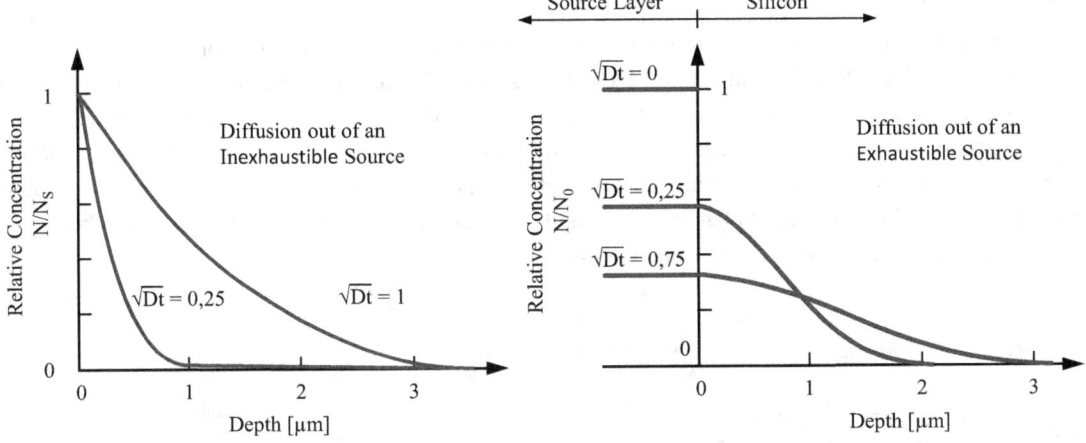

Fig. 2.23: Dopant profile for diffusion out of an inexhaustible and exhaustible source

During diffusion from an inexhaustible source, the saturation concentration of the dopant (N_S) in silicon is achieved at the surface. Since this concentration is too high for most electronic components, a diffusion process in an inert atmosphere usually follows. This leads to a distribution of the dopants in the silicon and a decrease in the surface concentration. The same effect is also achieved with diffusion from an exhaustible source. The initial concentration of the source layer (N_0) is distributed further and further in the silicon with increasing time.

Fig. 2.24 shows the dopant profile of a pin-diode, at which the p-doping region was diffused. The basis material is a high-doped n-silicon wafer, followed by a low-doped n-epitaxial layer. Into this epitaxial layer, Boron is diffused with a 100.000-times higher concentration as the n-basic dopant concentration of the epitaxy layer. Thus, a p-type area is produced, which is used as the anode of the diode.

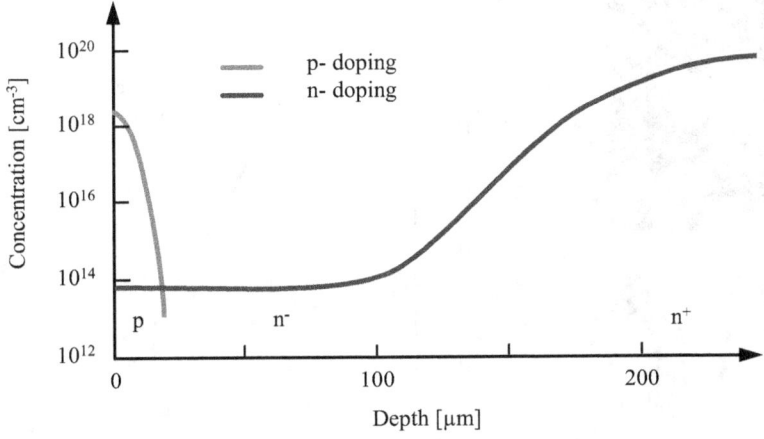

Fig. 2.24: Dopant profile of a pin-diode

Ion Implantation

Ion implantation is another way to bring dopants into the silicon. In this method, ions are accelerated by an electric field (10 to 200 keV) and shot on the silicon surface. During the penetration of the ions into the silicon lattice, they are slowed down, and produce a Gaussian-shaped distribution of dopants with a middle penetration depth of 0.1 to 2 μm. As seen in Fig. 2.27, the maximum doping level is not located on the surface. This can be used to create a buried doping area. At high energies and doses, it comes to an amorphization of the surfaces (Fig. 2.26). To re-crystallize this area and to incorporate the implanted dopants on a lattice site, a tempering process is necessary after the implantation.

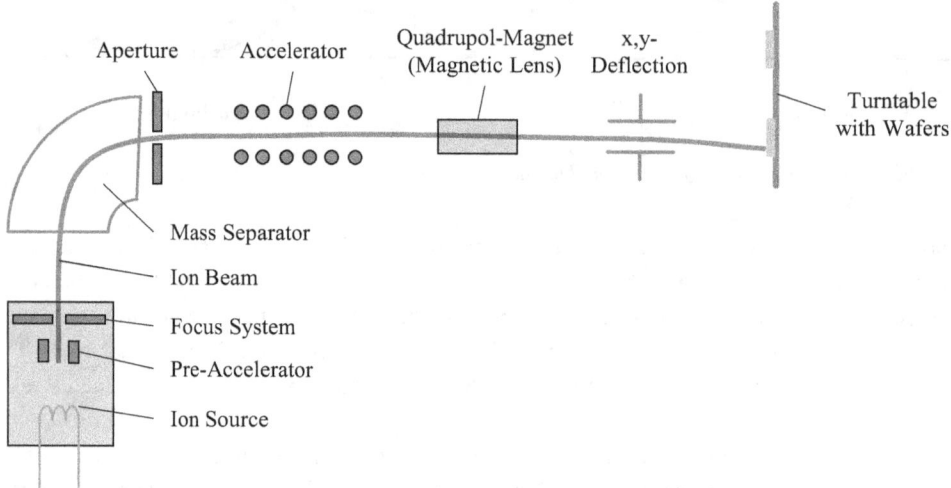

Fig. 2.25: Schematic representation of an ion implanter

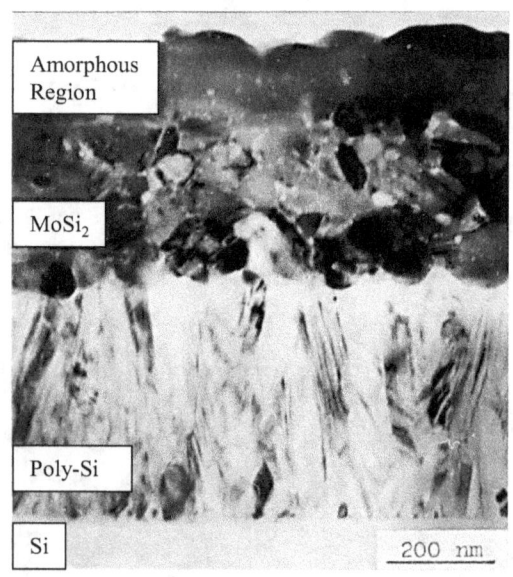

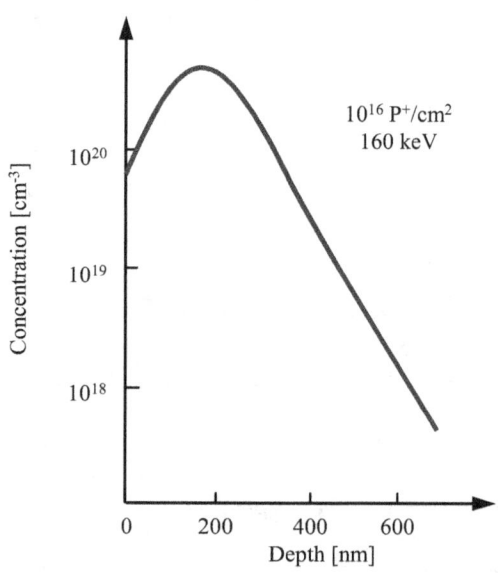

Fig. 2.26: SEM-cross section picture of an implanted $MoSi_2$-layer (Boron, $E = 40$ keV, $D = 5 \cdot 10^{14}$ cm^{-2})

Fig. 2.27: Dopant profile in {111}-Silicon, 7°-tilt (Phosphor, $E = 160$ keV, $D = 10^{16}$ cm^{-2})

Technology Example: Diffused Resistors Produced by a Planar Processes

The technology procedure of a diode discussed in the previous chapters, shows discrete electrical component manufacturing. In this example, only one component is realized on the silicon chip. However, the situation is quite different for integrated circuits. Here many components have to be placed on a silicon chip. To isolate the individual electrical components from each other, an isolation frame is placed around each component. Fig. 2.28 shows the structure of a bipolar OpAmp (µA741). The bipolar transistors, resistors and diodes are individually separated by an insulating frame. The electrical connection of the components in this example is realized by an aluminum wiring level.

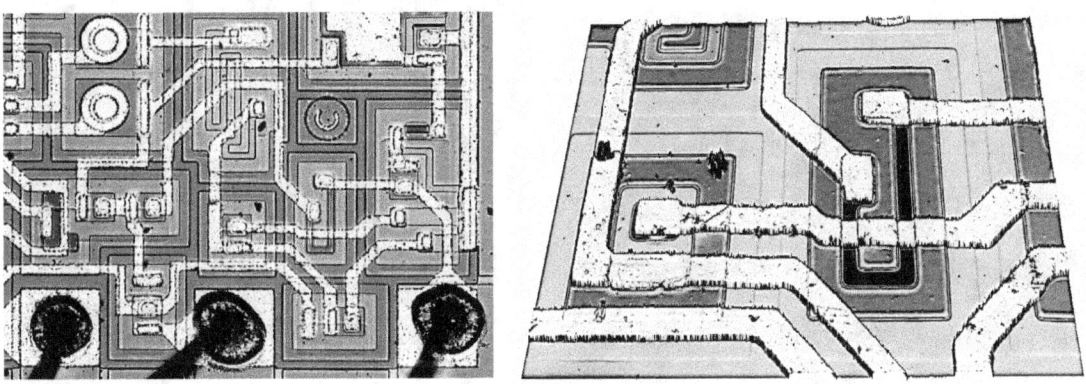

Fig. 2.28: Partial view of an integrated circuit (µA741) with isolating frame, and a detailed view of a diffused resistor

The following description shows the individual process steps for generating a diffused resistor, which is isolated by a p-doped frame from the rest of the silicon chip.

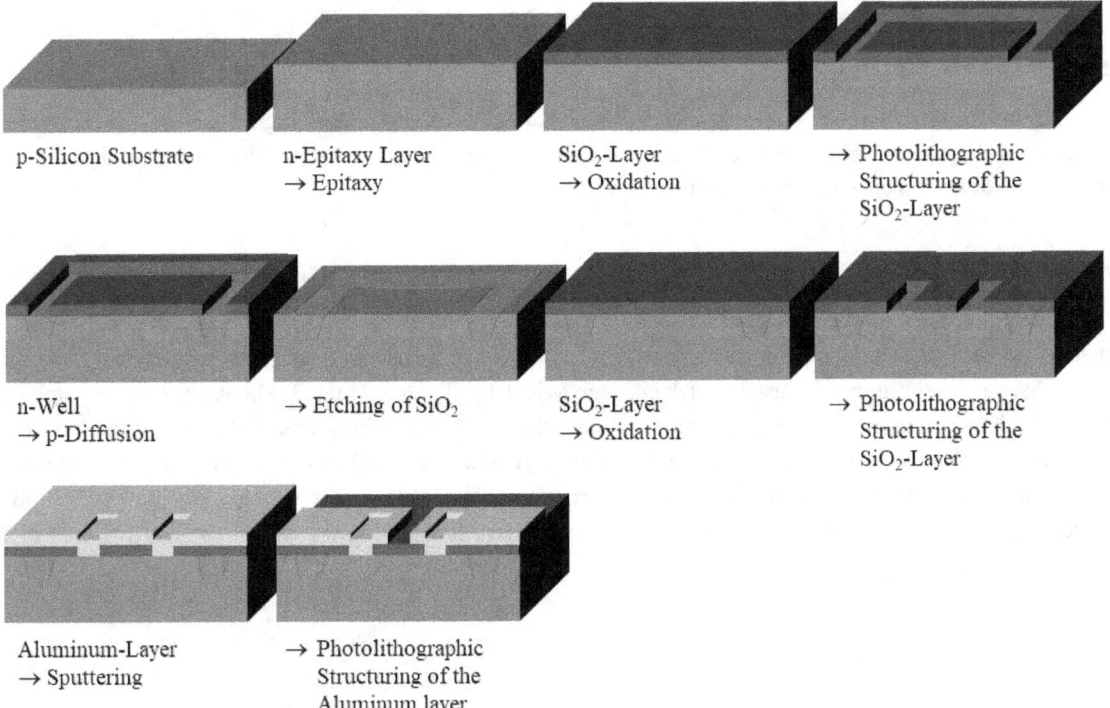

Fig. 2.29: Production process of a diffused resistor with isolating frame

2.3 MEMS Technologies

The technologies for the three-dimensional structuring of microsystems are divided into two categories: bulk, and surface micromachining. Fig. 2.30 and Fig. 2.31 show typical examples of each. For the bulk micromachining, deep structures are inserted into the silicon. These can reach depths up to the wafer thickness and can perforate the wafer. On the other hand, are only thin layers on the surface are structured for the surface micromachining, and the depths of the structure are a few micrometers. Both of these technologies coexist nowadays, but surface micromachining is used for the preparation of sensors significantly more in recent years.

Fig. 2.30: Technology example for bulk micromachining, spring-mass-system of an acceleration sensor

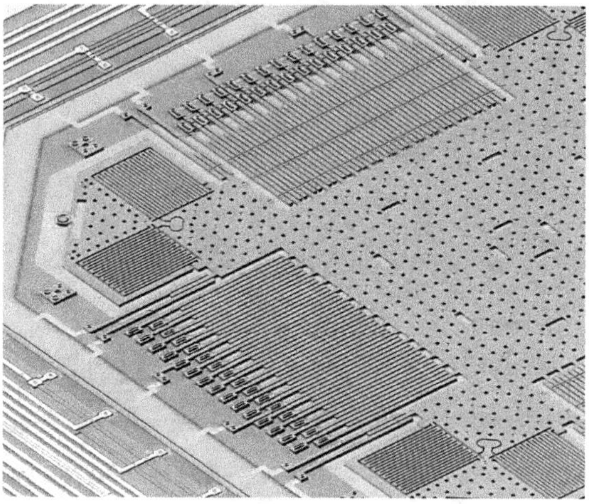

Fig. 2.31: Technology example for surface micromachining, structure of an acceleration sensor

Bulk Micromachining

For the deep structures of bulk micromachining, etching techniques with a high anisotropy are required. Orientation-dependent etching and reactive-ion etching are especially used as etching processes here. With them, it is possible to create deep and narrow trenches.

Orientation-Dependent Etching

Orientation-dependent etching can be used for etching crystalline silicon. It is based on the phenomenon that different orientations of silicon are removed with significantly different etching rates with certain etchants. Normally, aqueous solutions are used as etchant, such as KOH or TMAH. Fig. 2.32 shows the different etching rates of a KOH solution on {100}-silicon. It is seen, that at 0° the etching rate is 0 µm/min, whereas the maximum etching rate with 1.7 µm/min occurs in the range of 30°. Fig. 2.33 shows an etched test structure to determine the distribution of etch rates. The anisotropy of the etchant is easy to see on the structure in the middle of the wafer. Here the undercutting of the etch mask has produced a visual copy of the etch rate distribution.

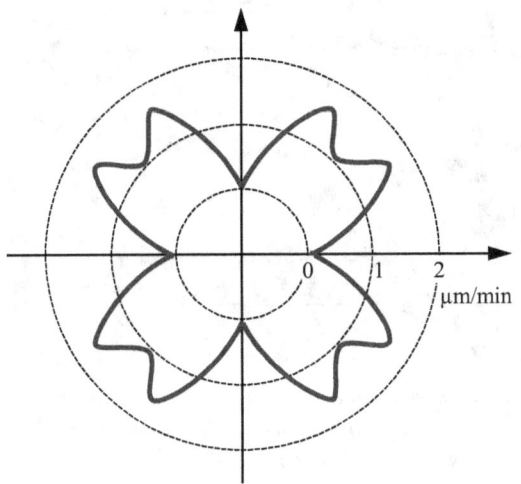

Fig. 2.32: Orientation dependent etch rates for a {100}-silicon wafer in aqueous KOH-solution (Concentration: 30 %; Temperature: 80 °C) with a <100>-depth etch rate of 1.1 μm/min

Fig. 2.33: Orientation dependent etched test structure on {100}-Silicon

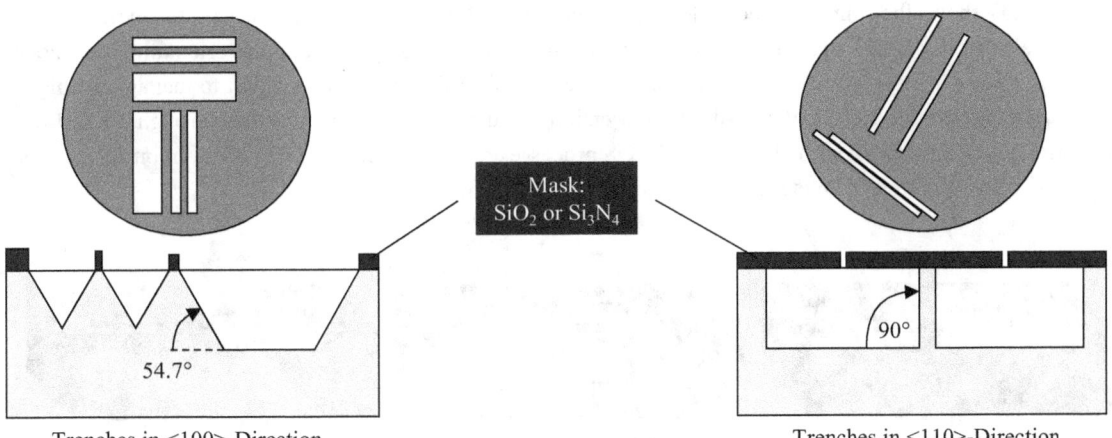

Fig. 2.34: Cross-section of etch profiles for different orientations of the mask openings on {100}-silicon

Depending of the wafer orientation and the mask edge directions, widely varying etch reliefs are produced. Two typical etched reliefs on {100}-silicon are shown in Fig. 2.34. At this wafer orientation, 54°-tilt {111}-etching walls are obtained for mask edges in 0° and 90°-direction, which are not attacked by the etchant. This means that such mask edges are not undercut. If the mask design is built only with mask edges in {100}-direction, a 1:1-copy of the mask structure is produced on the silicon surface. Convex corners are an exception in this mask design. Here, different crystal orientations occur, so that the corners are strongly undercut. To avoid this, convex corners are protected by so-called corners-compensation (see Fig. 2.35 and Fig. 2.36).

Another often-used configuration are mask edges in the 45°-direction. For this direction vertical sidewalls are seen for {100}-silicon. The disadvantage is that the sidewalls have no etch stop character. The etch rate is as same as the depth etch rate here and produce large undercutting of the etch mask.

 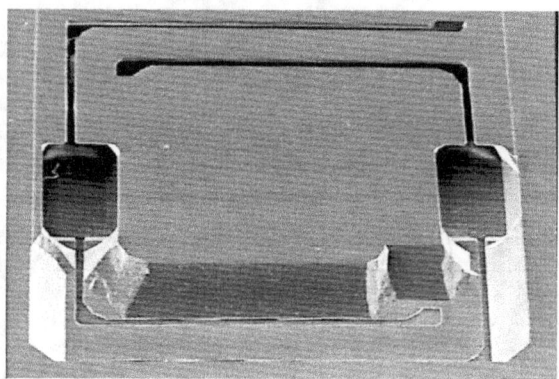

Fig. 2.35: Orientation-dependent etched spring-mass-structure with corners-compensations after 232 min etching time

Fig. 2.36: Orientation-dependent etched spring-mass-structure with corner compensations after 439 min etching time

Orientation-dependent etching is a very inexpensive procedure with a good reproducibility and low roughness on the sidewalls. It is used especially in the preparation of deep pits, and for the production of membranes for pressure sensors. Disadvantages of this method are that not all forms are possible, no isolated structures can be created, and the process has a bad compatibility to CMOS-processes.

Fig. 2.37 shows the simplified technology-flow for the preparation of a pressure sensor. It can be seen that first the electrically active structures along the top of the wafers are prepared, and the orientation-dependent etch process is carried out at the end. There are two reasons for this. First, it is difficult to manage the highly structured wafers after the orientation-dependent etching, and second, the wafer comes through a KOH-etch step strongly contaminated with alkali ions. CMOS processes react very sensitively to alkali impurities, so that it is hardly possible to feed an etched wafer back in a CMOS line.

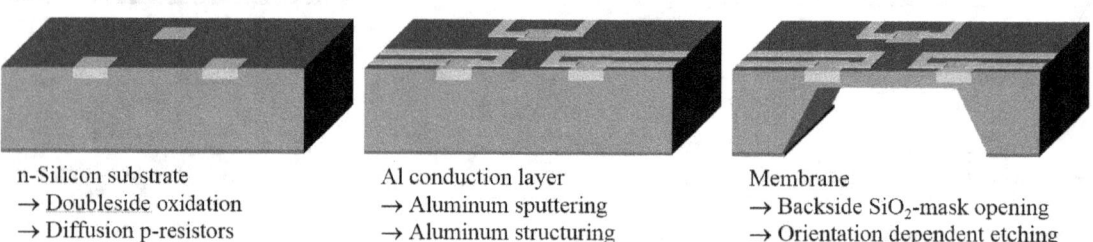

n-Silicon substrate
→ Doubleside oxidation
→ Diffusion p-resistors

Al conduction layer
→ Aluminum sputtering
→ Aluminum structuring

Membrane
→ Backside SiO$_2$-mask opening
→ Orientation dependent etching

Fig. 2.37: Process steps for producing a piezo-resistive pressure sensor using orientation dependent etching

Deep Reactive Ion Etching (DRIE)

The DRIE-process is the second possible procedure to etch deep trenches into the silicon. It is based on a sequential combination of ion etching and sidewall passivation. To accomplish this, a plasma is ignited alternately with SF_6 and C_4F_8, where SF_6 act as etchant and C_4F_8 as sidewall passivation. Since the SF_6 etching is anisotropic, the etching ground is attacked much more strongly than the sidewalls. To get an etch-trench with a high aspect ratio (depth divided by width of the etch trench) the passivation/etch step is repeated several times. This sequence of etching and subsequent passivation can be seen in the resulting wall structure, which is slightly wavy.

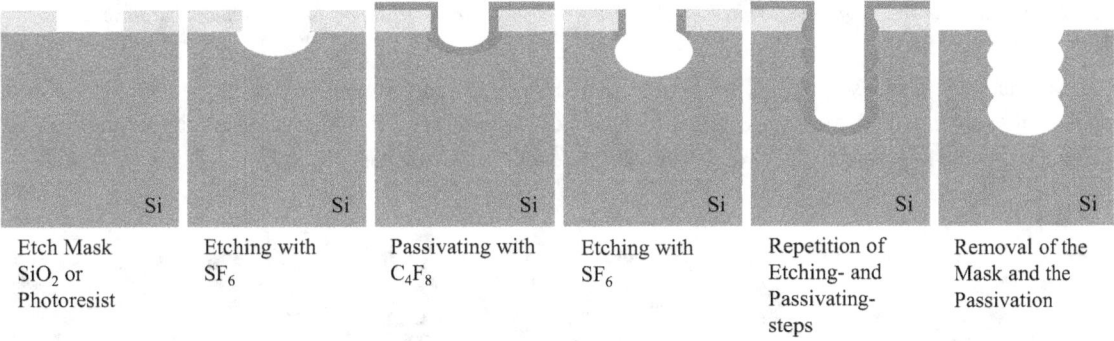

Fig. 2.38: Schematic representation of the DRIE-process

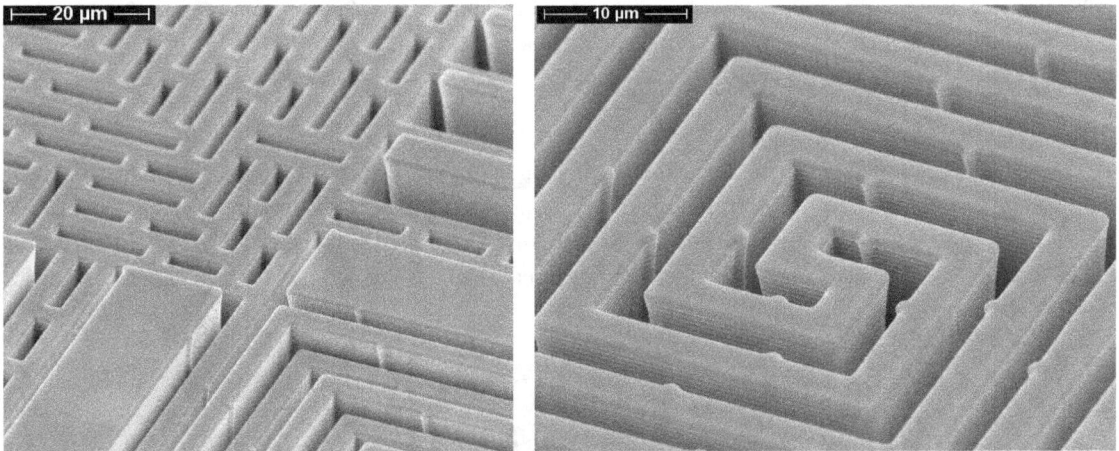

Fig. 2.39: DRIE-etched spring-mass-system with a depth of 40 µm

In most cases, the structures created by means of the DRI process are in a height range of 10 ... 100 µm. Since they do not perforate the wafer, special technologies must be used to cut out movable structures. Very often, silicon-on-insulator (SOI) technology is used for this purpose. Here, the micromechanical structure is created on a buried SiO_2 layer. After DRIE etching of the vertical trenches, the SiO_2 layer is locally removed by using an isotropic etchant (e.g., HF).

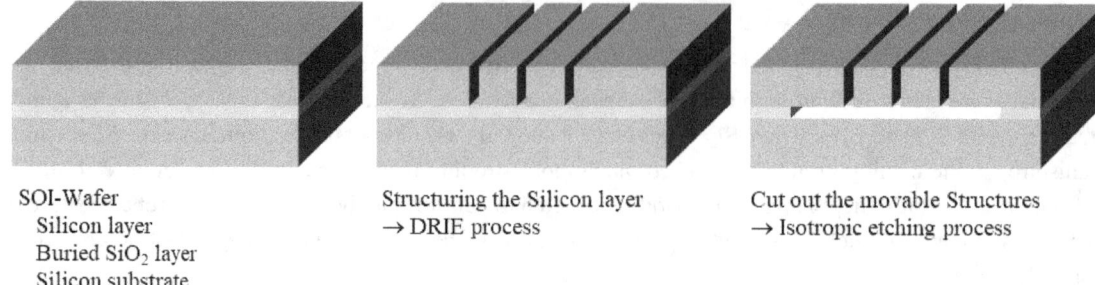

SOI-Wafer
  Silicon layer
  Buried SiO$_2$ layer
  Silicon substrate

Structuring the Silicon layer
  → DRIE process

Cut out the movable Structures
  → Isotropic etching process

Fig. 2.40: Schematic representation of micromechanical structures using SOI wafers to cut out movable structures

Alternatively, it is also possible to define the freestanding structures before the second silicon wafer is applied, thus saving the final etching step to free the moving structures. In addition to simplifying the process, it is also possible to define different depths under the functional elements.

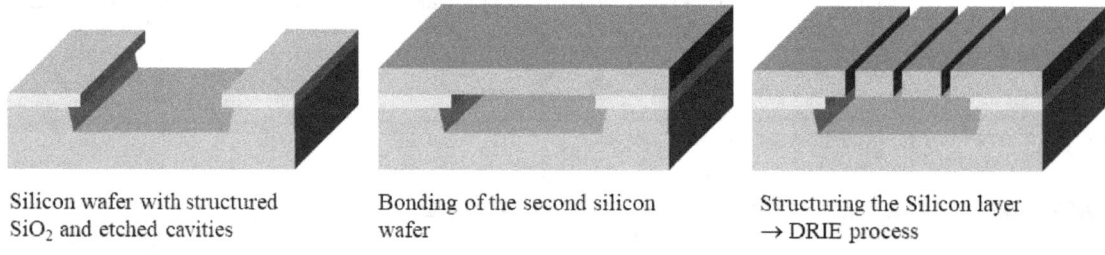

Silicon wafer with structured SiO$_2$ and etched cavities

Bonding of the second silicon wafer

Structuring the Silicon layer
  → DRIE process

Fig. 2.41: Schematic representation of the fabrication of micromechanical structures using pre-structured silicon wafers

The use of SOI wafers results in a very flat silicon structure that consists of single-crystal material, and thus has significantly better long-term mechanical properties than polysilicon structures. However, one disadvantage of SOI wafers is the relatively high cost involved in preparing them. Usually, two silicon wafers are bonded together for this propose, and then the functional side is thinned to the required dimension.

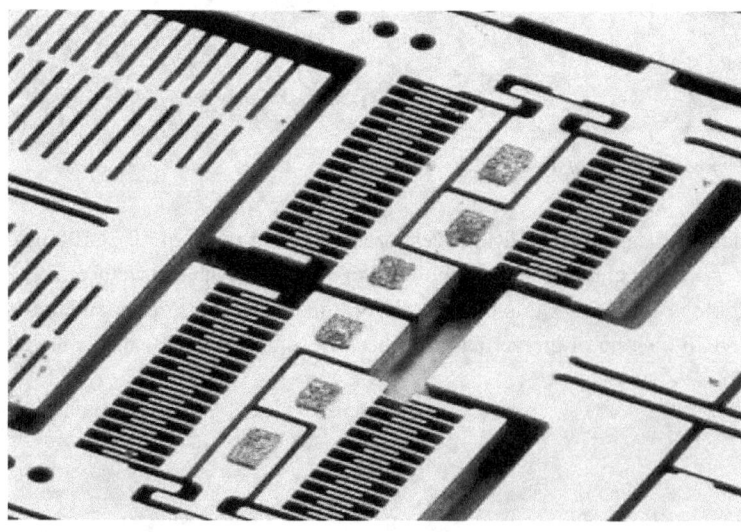

Fig. 2.42: MEMS structure on a SOI-Wafer

Surface Micromachining

Surface micromechanical sensors and actuators are mostly manufactured by using a sacrificial layer technique. For this technology, a polysilicon layer is deposited on top of a sacrificial layer (e.g., SiO_2) and afterwards the sacrificial layer is etched away. As a result a freestanding polysilicon layer is obtained, which can be used for movable mechanical functional elements. Fig. 2.43 describes the overall process of such a structure. A SiO_2-coated silicon wafers is used as substrate. On top of it a polysilicon conduction track is deposited, which can be optionally extended by further conduction layers. During the next technology step, the sacrificial layer is deposited by CVD and photolithographically patterned. Everywhere, where later a rigid connection between the function layer and the substrate should be, a window in the sacrificial layer is opened. It follows the deposition and the structuring of the polysilicon layer with a thickness of approximately 2 to 10 µm. As the last step of the process, the sacrificial layer is etched away. As a result, freestanding polysilicon structures are produced, which are fixed to the substrate only at the desired locations. As etchant, an isotropic etcher is used (e.g., HF-etching baths). The isotropic etchant produces an undercutting of the mask openings with the magnitude of the thickness of the sacrificial layer. By this, narrow structures can be completely set free. However, if larger connected areas occur, the sacrificial layer would not fully removed by undercutting. This is the case, for instance, for large mass elements of acceleration sensors. To ensure the freedom, many openings have to be integrated into the polysilicon layer (lattice structure). On the other hand, by omitting the etching holes, isolated and fixed structures can be obtained. This is important in the area of the anchor of the spring-mass structure in Fig. 2.43.

In addition to the combination of SiO_2 and polysilicon, there are other combinations of materials for the realization of sacrificial layer systems, but this subject will not be dealt with in detail here.

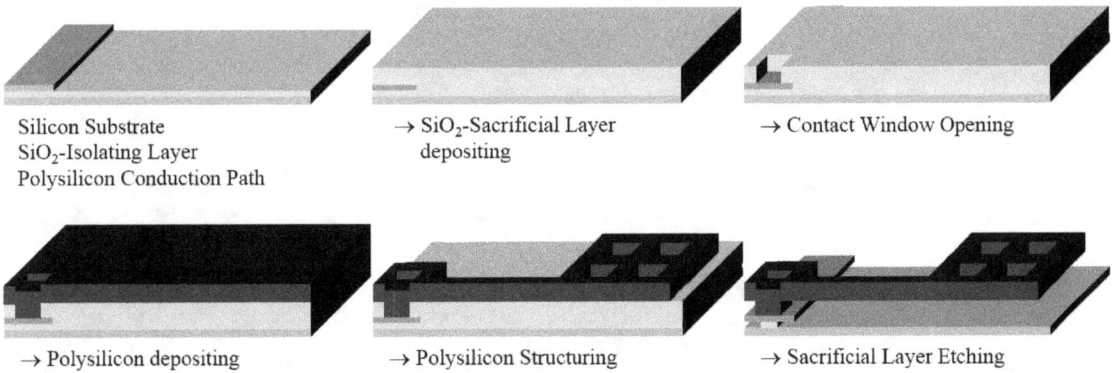

Fig. 2.43: Schematic representation of the sacrificial layer process

Fig. 2.44 shows the spring structure of an acceleration sensor, which was produced using the sacrificial layer technique. It contains the anchor structure on the left and the freestanding spring structure on the right. To ensure that the spring is completely undercut, two elongated openings are inserted here. This figure clearly shows the undercut of the polysilicon layer in the area of the anchor structure. Although it results in a tapering of the anchor structure in this area, it is not a complete undercut.

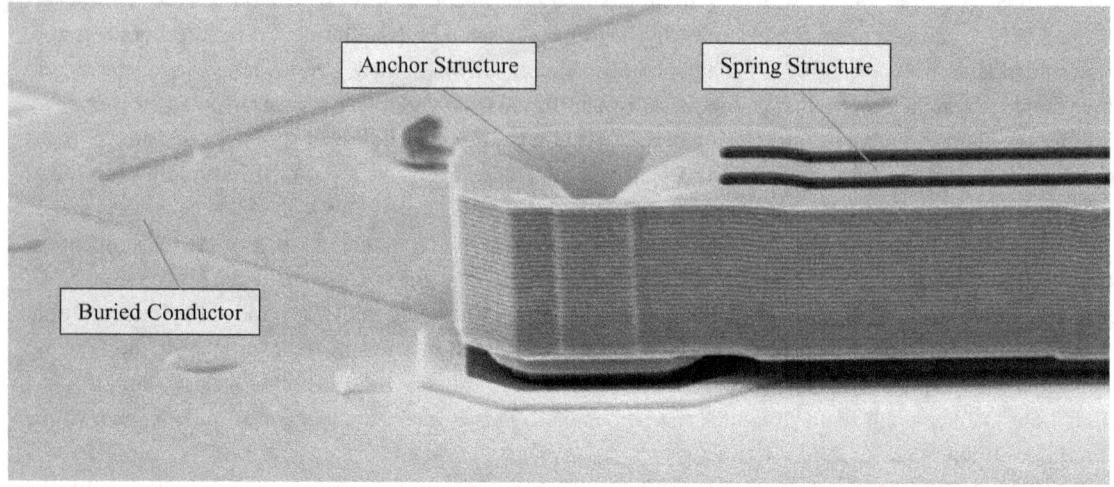

Fig. 2.44: Spring structures of an acceleration sensor produced using sacrificial layer technology

Wafer Level Packaging

Packaging of microsensors differs significantly from microelectronics, because the generated sensor structures need a protected cavity. This is necessary in order to ensure the freedom of movement of the mechanical structures, and to protect the microsystem from dust and humidity. The hermetic closure has to happen before the individual sensor is separated from the wafer, as the separation process (sawing with cooling liquid) would destroy the unprotected microstructures. Fig. 2.45 shows a possible implementation variation of a surface-micromachined system. The silicon cover is soldered by means of solder glass on the microstructure and seals the interior hermetically.

After the microsystem is fully protected by the wafer-level packaging, the individual sensors are sawn out from the wafer and connected by the methods known from microelectronics. Afterwards it is housed in an IC-package. Fig. 2.46 shows the complete packaging of the acceleration sensor ADXL345.

Fig. 2.45: Wafer-Level-Packaging

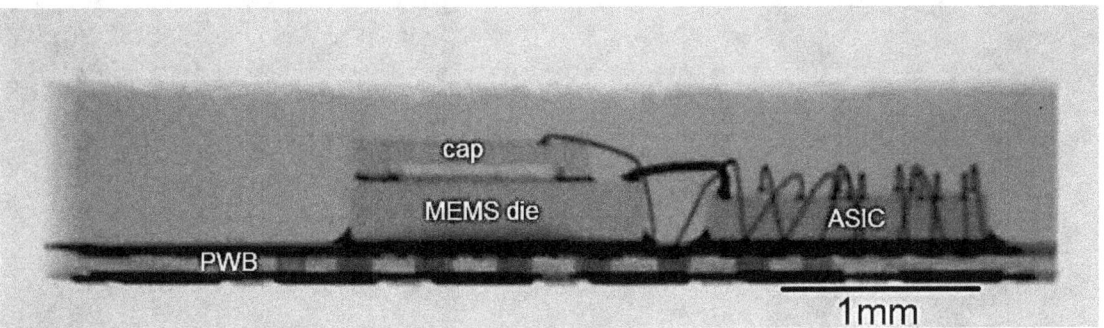

Fig. 2.46: Packaging of the acceration sensor ADXL345

(Source: Analog Devices)

For fully integrated microsystems, in which both the electronics and sensor structure are integrated on one chip, other technology solutions are generally used for wafer-level packaging. Here, SiTime's process for generating resonator structures is presented as an example.

The starting material is an SOI wafer in which the resonator structure is created by DRIE etching. In the first step, the entire surface, including the etched trenches, is coated with SiO_2. Only the areas of the electrical contacts are cut out again. In the next step, a thin polysilicon layer is deposited and fine holes are structured above the resonator structures. The etchant (HF-vapor) can then penetrate through these holes and etch free the trenches filled with SiO_2. The following polysilicon deposition fills the narrow holes again and seals the cavity hermetically. To contact the resonator structure, the contact windows are cut out in the following process step and the trenches are filled with SiO_2 to isolate the contacts from the rest of the silicon wafer.

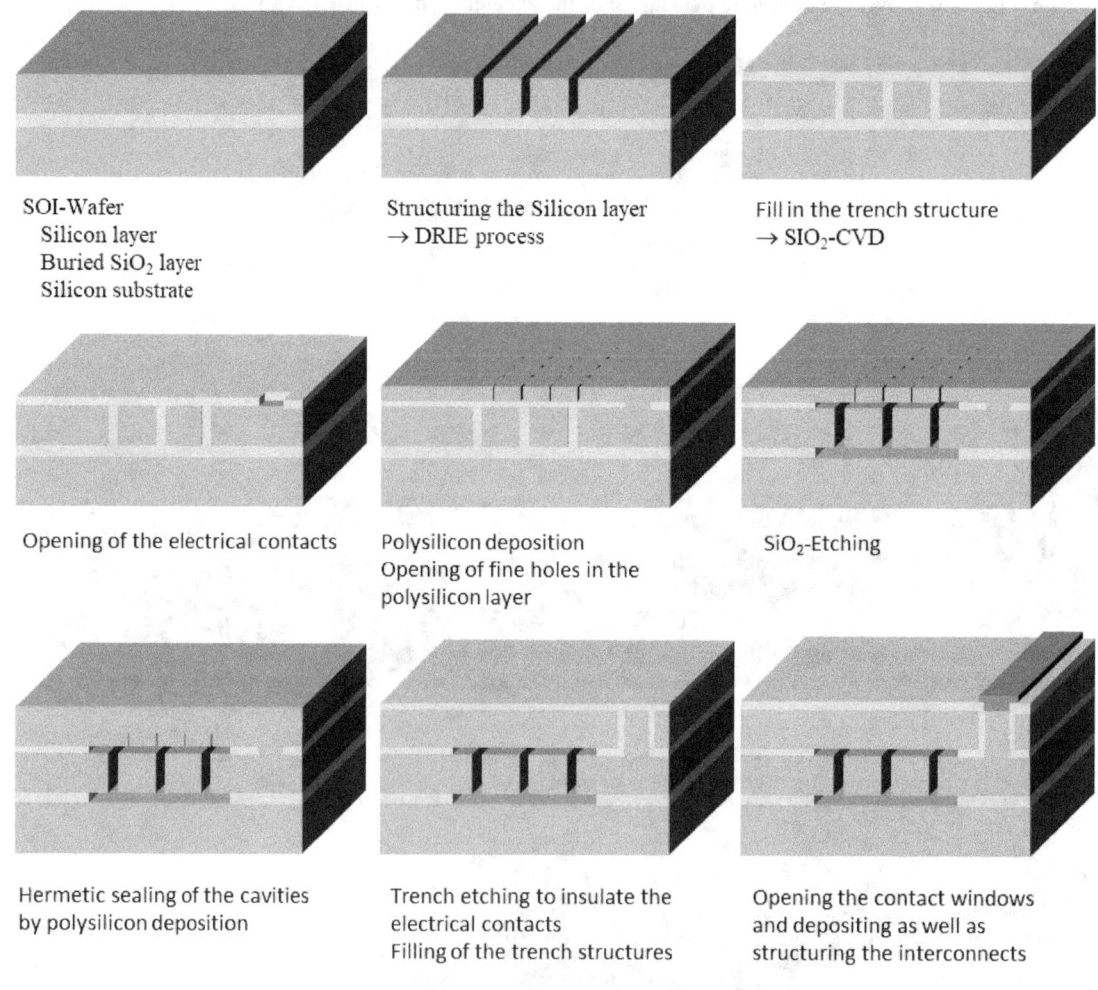

Fig. 2.47: Schematic representation of the process of SiTime for the encapsulation of micromechanical structures

2.4 Packaging

The packaging of microsystems encompasses several levels. After the wafer-level packaging, which was described in the previous chapter, most microsystems are installed on a circuit carrier, such as a commercial PCB. To be able to use the standard solder processes, the microsystem is packaged into an IC-package. At the end, the entire system with sensor, actuator and evaluation electronics, is mounted into a housing, to protect it against environmental influences.

Housings for electrical and electronic devices are classified according to the IP-code of the standard IEC 60529. The first digit of the code represents the protection against foreign objects and touching, and the second digit indicates the protection against water. Therefore, a housing with a classification of IP68 is dust and waterproof. Industrial installations often use systems with a protection classification of IP54. These housings protect against accidental contacts, and mostly against dust and splashing water.

X	Protection against foreign objects	Protection against touching
0	No protection	No protection
1	Protection against foreign objects with a diameter > 50 mm (2 inch)	Protection against touching with the back of the hand
2	Diameter > 12.5 mm (0.5 inch)	Protection against touching with a finger
3	Diameter > 2.5 mm (0.1 inch)	Protection against touching with a tool
4	Diameter > 1 mm (0.04 inch)	Protection against touching with a wire
5	Protection against dust in harmful quantities	Complete protection against touching
6	Dust-proof	Complete protection against touching

Tab. 2.4: Meaning of the first digit of the IP-Codes according to DIN EN 60529

X	Protection against water
0	No protection
1	Protected against dripping water
2	Protected against dripping water, in case of the housing is tilted by max. 15°
3	Protected against spray water, in case of the water comes at a max. 60° tilt from the vertical line
4	Protected against spray water from all directions
5	Protected against jet water from all directions
6	Protected against heavy jet water
7	Protected against temporary submersion under water
8	Protected against continuous submersion
9	Protected against water from high-pressure or steam cleaning

Tab. 2.5: Meaning of the second digit of the IP-Codes according to DIN EN 60529

IC Packages

IC packages are mainly used for the packaging of microsystems. This is always possible, if the microsystem requires no special connections for the supply of media and other measured variables.

According to the choice of packaging material, housings are divided into plastic and ceramic packages. Most of the package are offered both as plastic and as ceramic package. Usually, the type of material is specified as the first letter. There are for example LCC packages with a plastic body (PLCC) and a ceramic body (CLCC). Plastic enclosures are the cheaper alternative, but have significantly worse parameters than ceramic packages. This can be seen especially in the thermal expansion coefficient and of the tightness to water vapor.

On all packages, Pin1 is marked. This is done by a notch or circular depression in the housing, a flattening of a corner or edge of the housing or by printing a number or a point on the housing. The pins are arranged in the counterclockwise direction seen from the top of the package. The pin spacing (grid) has shrunk significantly in recent years. The first DIL-package had a pitch of 2.54 mm (0.1 inch). Today, the finest grids are in the range of 0.3 mm (0.012 inch). It can also be observed in recent years, that circuits with metric and inch-grid will be offered. The exact pitches and the recommended bond pad geometries are provided by every manufacturer in the data sheet.

The first package for integrated circuits was the dual-in-line housing. It has two rows of through-hole contacts. The pin spacing is 2.54 mm (0.1 inch). DIL packages are nowadays rarely used due to their large area consumption. However, for prototyping and for applications in which the circuits are socketed, they are an inexpensive alternative to other types of packages.

Fig. 2.48: DIL package (through hole technology)

Since the 1980s, the DIL packages have been replaced by small outline (SO) packages. They have the advantage of being compatible with SMD technology. Furthermore, the grid and the height of this package were reduced step by step. A reduction in the assembly area was achieved with the SOJ packages, in which the pins are bent under the package. However, the bent pins increase the height of the mounting and complicate the optical quality control of the solder joints. That is why SOJ-packages are only sporadically in use today. In order to further increase the possible pin count compared to SO-packages, packages with pins on all four sides (QFP).

SOP (Small Outline Package)	SOJ (Small Outline IC "Gull Wing" Style)	QFP (Quad Flat Package)

Fig. 2.49: SO and QF packages

Through further miniaturization of package shapes, the "no-lead" packages were introduced. In these packages, the solder pins were omitted, and only the connection surfaces were attached below the housing. On the one hand, this leads to the saving of the contact surfaces and to a further reduction in the package height. The major disadvantage of this design is that the soldering points can no longer be visually inspected since they are located under the component.

Fig. 2.50: QFN and LGA packages

The QFN package has metal plates as connection surfaces, which are punched out of a frame. After contacting the silicon chip by means of wire bonding, the entire circuit is encased in plastic encapsulation. Due to the punched metal connection plates, the free arrangement of the solder pads is only possible to a very limited extent. On the other hand, it is possible to achive active thermal cooling in addition to electrical contact via the terminals. For this reason, it is not uncommon to find a ground plate in the middle of a QFN package, to which a heat sink structure can be connected.

LGA packages (land grid array) are visually similar to QFN packackes, and all connections are also located below the circuit. Since a printed circuit board is used as the base plate, the connection geometry can be freely selected. The printed circuit board (interposer) consists of a BT-substrate (see Tab. 2.6) with two or four layers and gold-plated copper conductors. Here, contact to the silicon chip is also made by means of wire bonding.

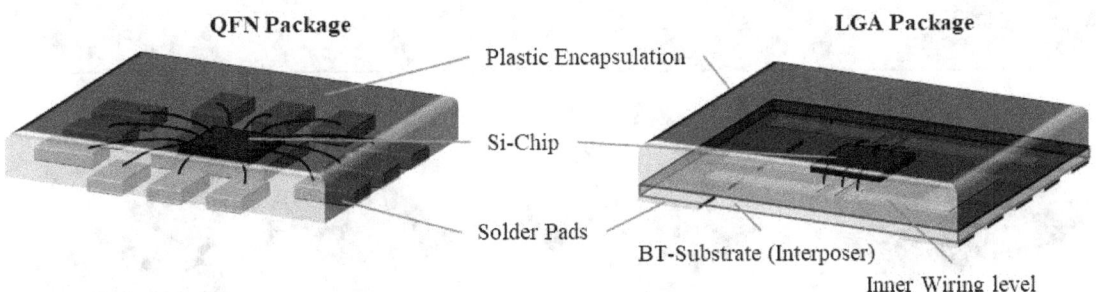

Fig. 2.51: Structure of QFN and LGA packages

MEMS

Chip-scale packages (CSP) offer a further increase in integration density. Their connection pads are integrated on the silicon chip directly. For mounting on the circuit board, the bare silicon chip is soldered with the component side down (flip chip). In many cases the connections on the silicon chip are equipped with solder beads (ball grids) to make the process technologically easier.

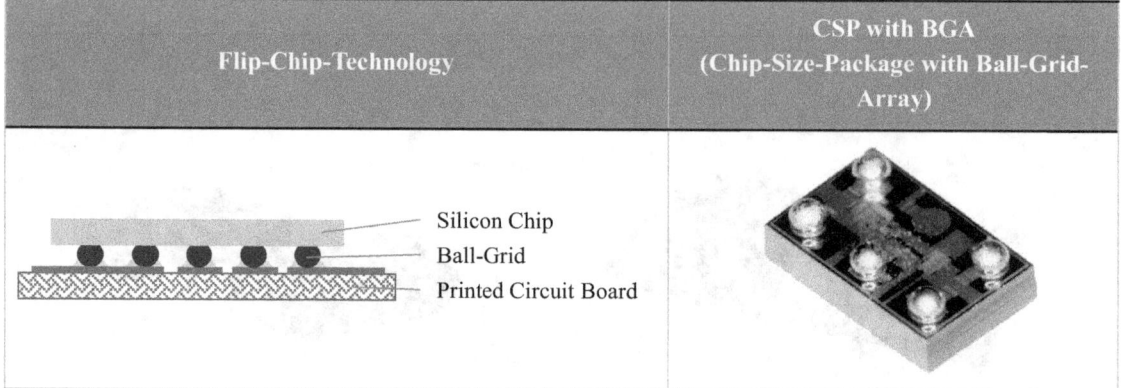

Fig. 2.52: Flip-Chip-Technology and Chip-Size-Package

MEMS Packages

Practically all package types described in the previous section can also be used for microsystems. Primarily, plastic-molded LGA packages are used due to their low construction volume and low package costs. Inertial sensors (acceleration sensors and gyros) can be packaged using standard LGA packages. They do not require any mechanical connection to the environment and can therefore be fully encapsulated. The situation is different for pressure sensors, for example. For these, it is necessary to bring the medium, in which the pressure is to be measured, in contact with the silicon chip. Either the silicon chip is allowed to peek out on one side of the package or a package with a built-in cavity is used.

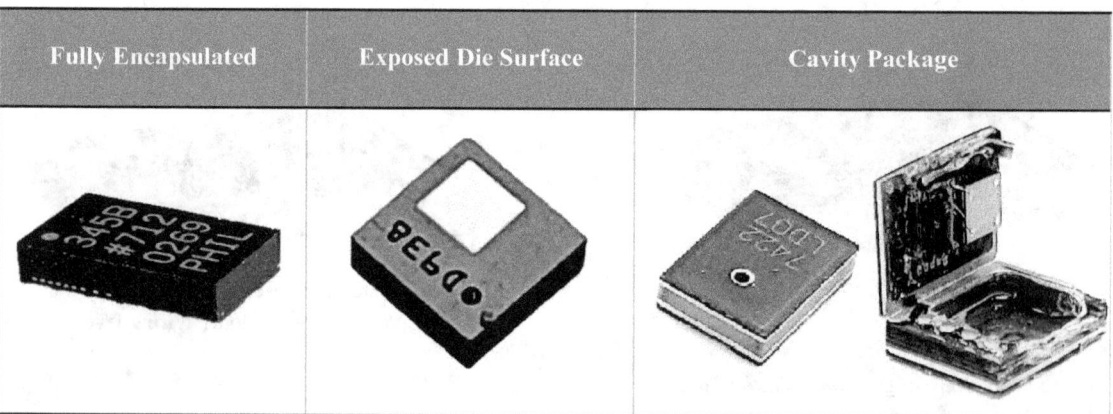

Fig. 2.53: Types of LGA packages for MEMS

For cavity packages, ceramic is often used as material. The advantage of using ceramic packages in favor of plastic molded packages is that they are gas-tight. This makes it possible to protect the microsystem against the effects of moisture from the environment or to provide a defined protective gas atmosphere (N_2, SF_6) in the cavity.

Fig. 2.54: CLCC package of an acceleration sensor

In some cases, it is necessary to enable a physical coupling of media or influencing variables to the microsystem. Since this is not possible with conventional packages, IC packages are adapted, or their own housing shapes are developed.

Fig. 2.55: Pressure sensor with pressure inlet

Fig. 2.56: Micro mirror array with quartz window

Fig. 2.57: Micro pump with in- and outlet

Wire Bonding

The electrical connection of silicon chips is usually carried out by means of bonding connections. Two different materials are used for bonding wires: gold and aluminum.

Fig. 2.58: Gold wire processed by thermosonic bonding   *Fig. 2.59: Aluminum wire processed by ultrasonic bonding*

Aluminum wires are processed by means of the ultrasonic bonding procedure. The aluminum wire is pressed on the bonding pad, and welded by using ultrasound. During the process, the wire is strongly deformed, and wedge-shaped structures are created. For this reason, the procedure known as wedge bonding.

The connection of gold wires may be done in the same way. Alternatively, for gold another method exists, the thermosonic process. With this method, the gold wire is melted before the bond operation and a gold ball at the end of the wire arises. After the ball has solidified, it is pressed onto the bonding pad and welded by means of ultrasonic vibration. The other side of the bond connection is then contacted again, as in the ultrasonic bonding process. One advantage of the thermosonic process is the vertical exit angle of the wire from the spherical bond point. Since it has no pinched constrictions, the bond connection is much more reliable and is well suited to bridging large differences in height between the chip and the package.

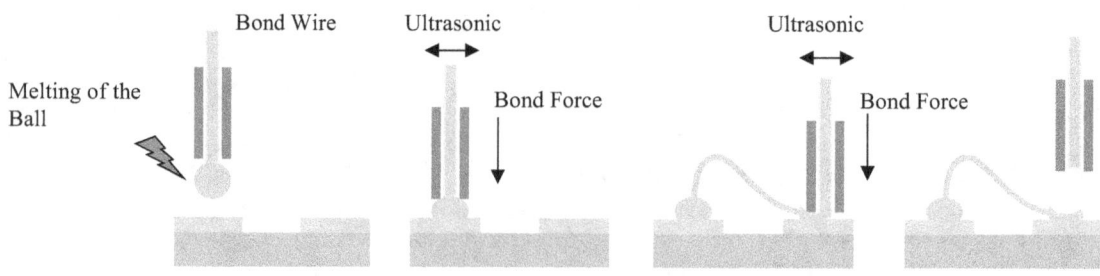

Fig. 2.60: Schematic representation of the thermosonic bonding procedure

Printed Circuit Board Technology

Printed circuit boards are the most commonly used circuit substrate in microelectronics and microsystems technology. They were developed in the 1950s, and constantly adjusted in complexity to the present day. Double-sided boards with through holes (vias) represent the basic configuration of the circuit board. They consist of a base material, which is laminated on both sides with copper foil. To implement the electrical connections from the front to the back, the layer stack is drilled, and then the edges of the hole are electroplated with copper.

Industrially manufactured PCBs also consist of a solder resist and a protective layer for pads. The solder resist prevents molten solder paste from spreading widely during soldering. Furthermore, the protective layer supports the solder process by preventing the oxidation of the copper and ensures a good wetting of the molten solder.

Fig. 2.61: Structure of a double-sided PCB with through holes

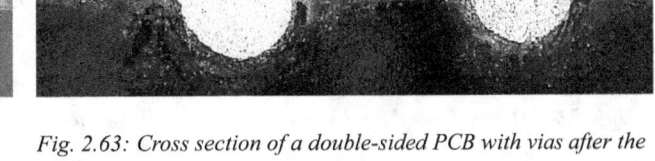

Fig. 2.62: Top view of a gold plated PCB with vias

Fig. 2.63: Cross section of a double-sided PCB with vias after the soldering process

Tab. 2.6 shows the most common materials and their parameters in PCB production. For the name of the base material, the FRx-nomenclature is often used. So-called FR4 is a glass fiber reinforced epoxy substrate. It is the standard base material for industrial use, and is stable up to 120 °C operating temperatures. For higher temperatures, which can occur for example in an engine compartment, FR5 boards or polyimide/glass substrates are used. On the other hand, in the low-cost consumer sector FR2/3 circuit boards are often found. They consist of a paper-reinforced epoxy or a phenolic resin mixture, and are significantly cheaper in their processing. For simple applications, single-sided PCBs are frequently used, in order to save the cost of the chemical through-hole plating.

Rolled foils made of electrolytic copper (99.9% Cu) are processed for the copper conductor tracks. The standard thickness is 35 µm. For high-current applications, greater layer thicknesses are used, up to a thickness of 3 mm. Since the large layer thicknesses would strongly increase the possible minimum feature sizes, the copper layers are galvanically reinforced after the patterning in this case. On the other hand, for very fine structured PCBs, very thin copper layer thicknesses until a minimum thickness size of 5 microns are used.

Solder Resist	Photoresist or Foil	d = approx. 40 µm	
Protection Layer	Chemical Sn	$d > 1$ µm	Soldering Press-fit technology
	Hot Air Leveling (HAL)	$d > 5$ µm	
	Chem. Ni/Au (Flash)	$d_{Ni} = 3 - 6$ µm $d_{Au} < 0.1$ µm	Soldering Wire bonding, conductive gluing
	Galvanic Ni/Au (electroplated hard gold)	$d_{Ni} = 4 - 6$ µm $d_{Au} = 1 - 5$ µm	Contacts
	Organic passivation	$d = 0.5$ µm	Soldering
Copper Layer	99.9 Cu	d = 5, 9, 18, **35**, 70 µm to 3 mm	
Base Material	FR4 Epoxy/Glass $d = 0.5 - $ **1.6** $ - 3$ mm	Standard for industrial applications	
	FR3 Epoxy/Paper FR2 Phenolic Resin/Paper	Cheap and disposable electronics, partly in large domestic electrical equipment and home electronics	
	FR5 Epoxy/Glass PIstarr Polyimide/Glass	Applications with higher temperature requirements $T_g = 160$ °C, for permanent use until 140 °C $T_g = 260$ °C, usually use until 200 °C	
	FR5 BT Bismaleimide-Triazine-Epoxy /Glass	Interposer for circuit packages $T_g = 180$ °C, good chemical and thermal resistance even with multiple soldering processes	

Tab. 2.6: *Materials used for the functional layers of a printed circuit board (FRx - designation from the American standard)*

Protective layers are usually made by chemically deposited tin or gold layers. These have the advantage that they grow very flat and evenly on the copper pad, compared to HAL-deposited tin layers. Tin coatings have layer thicknesses of 1-5 microns. Gold layers are significantly thinner coated. On the one hand, this is done for costs reasons, but on the other hand, it also has a technical reason. Gold dissolves during the soldering process in the solder and generates intermetallic gold-tin compounds, which can weaken the solder joint. The less gold is on the solder pad, the less it affects the solder joint. However, a thin layer of gold cannot fully protect the underlying copper layer. Therefore, an additional nickel intermediate layer is deposited as a diffusion barrier. For plug-in contacts and sliding contacts, the thin gold layer of flash gold plating is not enough. For this application, a much stronger gold plating is used.

Fig. 2.64 shows the separate technology steps for structuring a double-layer printed circuit board. After drilling and before the copper etching process, the vias are catalytically metallized. If necessary, the metallization of the vias is subsequently reinforced electrolytically. Since this process is not selective, copper is also deposited on the top and bottom of the PCB.

Various etchants are used for the Cu etching process. Frequently, an aqueous solution of iron(III)chloride ($FeCl_3$) or ammonium persulfate ($(NH_4)_2S_2O_8$) is used. The most important factor is a good selectivity of the etchant with respect to the etching mask. The photoresist layer used for pattern transfer is usually not sufficiently stable for the etching process. Therefore, after photolithographic patterning of the photoresist, the then uncovered copper surfaces are electroplated with a tin layer. After removal of the photoresist, this tin layer acts as an actual etching mask for the structuring of the underlying copper. The tin layer is removed after etching because it is heavily contaminated and attacked by the copper etcher, so that it can no longer be used as a protective layer for the soldering process.

MEMS

1. Copper-laminated base material

2. Drilling of vias with a diameter down to 0.3 mm
3. Galvanizing of the vias

4. Photoresist applying
5. Photoresist exposure with UV-light
6. Photoresist developing and wash out of the exposed areas

7. Open Cu-areas galvanizing (Sn)

8. Photoresist removing

9. Cu-etching
10. Sn-stripping
11. Optional galvanizing of addition Cu-layer

12. Applying and structuring of the solder resist

13. Applying of protection layers on open Cu-areas

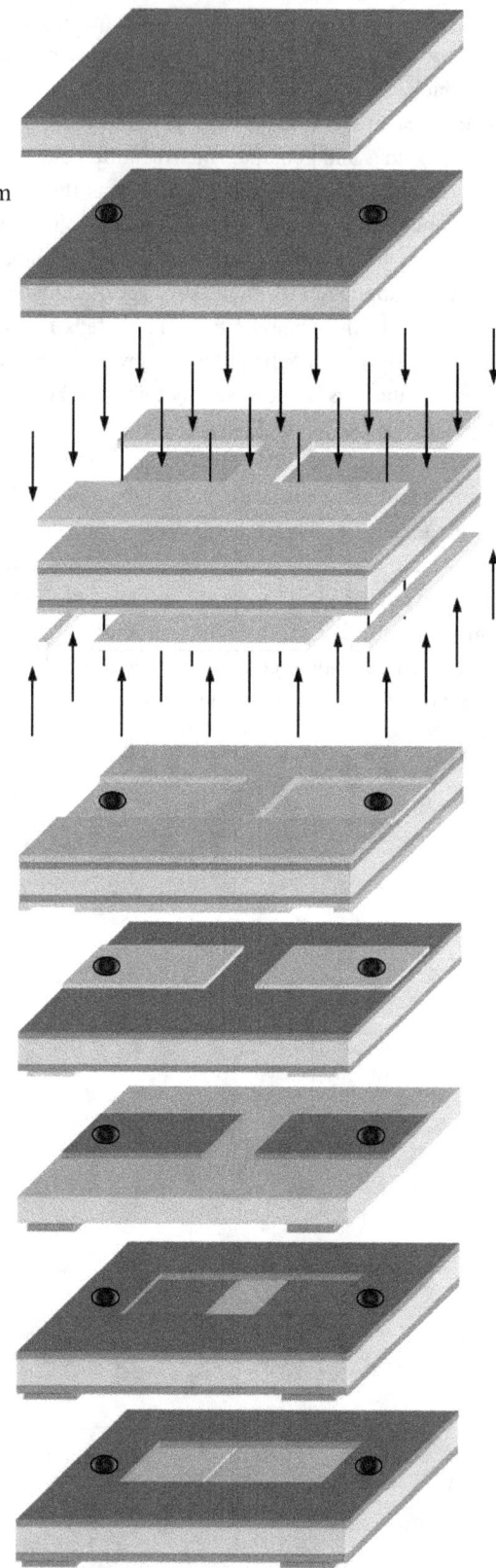

Fig. 2.64: Producion steps of a commercial double layered PCB with vias

Multilayer printed circuit boards are used for complex board designs. They include mostly six copper layers, which can be fully connected between with each other. Fig. 2.65 shows the structure of such a circuit board. The manufacturing process of a multilayer PCB is similar at the beginning to double-layered PCB-manufacturing. After structuring the copper layer, a further layer is applied on both sides, which consists of a non-polymerized carrier layer (Prepreg). In the next step, a copper layer is laminated, and then structured. At the end of the production process, the entire layer stack is polymerized at high temperatures and pressures. With this technology, printed circuit boards can be produced with up to 24 copper layers.

Fig. 2.65: Structure of a 6-layered PCB with vias

Furthermore, it is also possible to use flexible carrier materials. In this manner rigid-flex constructions are produced, which are suitable for 3-dimensional designs of PCBs in special housing forms.

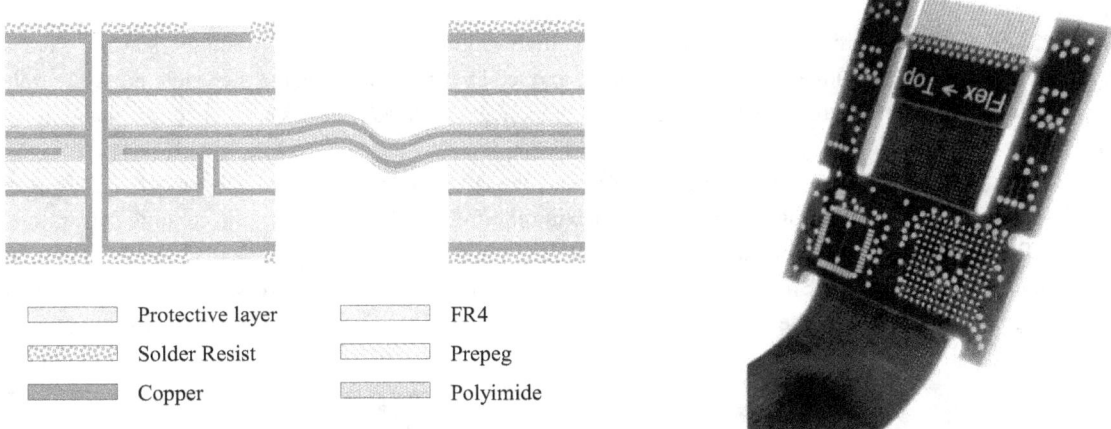

Fig. 2.66: Structure of printed circuit boards with rigid and flex-elements

Thick-Film Technology

The thick-film technology is an additive process for producing circuit carriers. For it, ceramics (Al_2O_3, AlN) are used as base material. Pastes are applied by screen-printing, which forms the electrically active components after a sintering process. In addition to conductive structures, it is also possible with this technology to make non-conductive layers as well as resistors and capacitors. Al_2O_3 has significantly improved electrical and thermal properties than FR4 material. For this reason, the thick-film technology is often used for applications with extended temperature range.

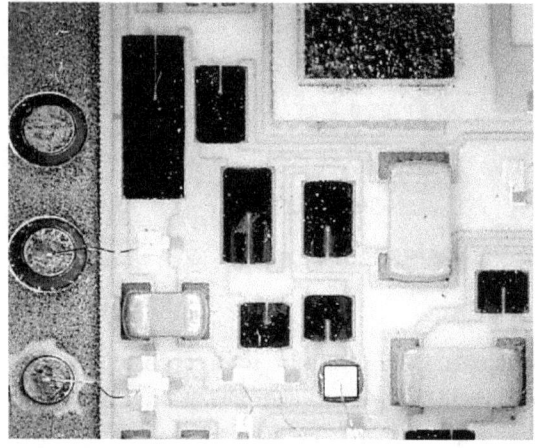

Fig. 2.67: Thick-film circuit carrier with laser-trimmed resistors

Fig. 2.68 shows the individual technological steps involved in the thick-film technology. Starting from a ceramic substrate, the individual functional layers are applied by screen-printing and then sintered. The result is a multilayer arrangement that can contain passive components, such as resistors and capacitors, in addition to conductors and insulation layers.

In thick-film technology, it is not necessary to contact the layers with each other by means of vias, as used in printed circuit board technology. However, connecting the front and backsides of the ceramic substrate is more difficult. Here, laser ablation can be used to drill holes in the ceramic substrate, which are filled with a conductive paste and sintered. Since this technique is very expensive and prone to errors, soldered contact clips are often placed on the side of the substrate to establish the electrical connections between the front and back.

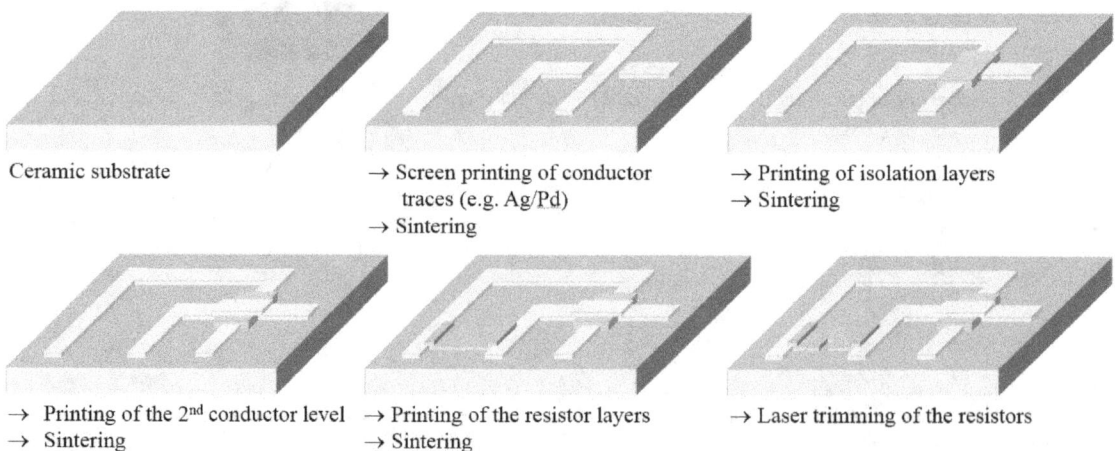

Fig. 2.68: Production steps of a multi-layered ceramic circuit board (MLC)

Soldering

Soldering is a thermal process for the integral joining of materials. Here one partner of the soldering process is melted (e.g., tin) whereas the second partner is kept well below its melting temperature (e.g., copper). During the soldering process, a surface alloy is formed on the contact surface of the soldering partners through diffusion processes. Soldering processes differ from welding processes in that they have lower process temperatures, since both process partners do not have to be heated up to their melting temperature. A distinction is made between hard and soft soldering. Solders with melting points above 450 °C are referred to as hard solders, below as soft solders.

When soldering electronic assemblies, a tin alloy is melted and an alloy is formed with the respective soldering partner. When soldering on copper, tin-bronze layers that are micrometers thick are created. It should be noted that for this alloy formation, copper of the conductive layer is consumed, which can lead to the complete dissolution of the copper layer when soldering thin layers or when soldering several times (e.g., for repair processes). If this is to be avoided, diffusion barrier layers are used. Fig. 2.69 shows this for the contact surface of a resistor. In this case, the thin silver-palladium contact surface is protected by a nickel layer.

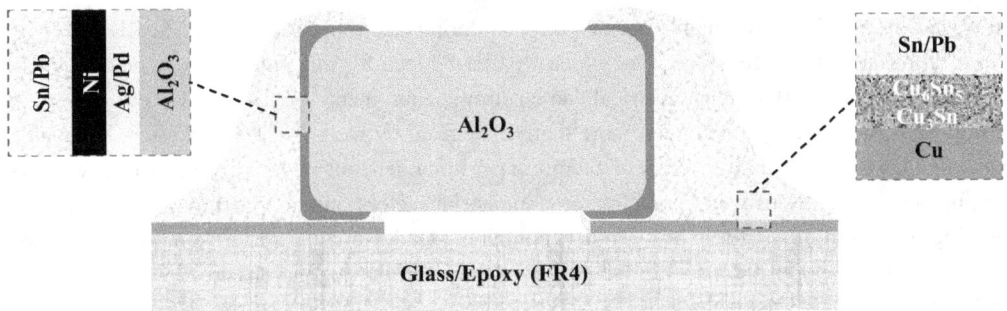

Fig. 2.69: Soldering of a SMD component at a PCB

During soldering, a flux is used to ensure better wetting of the molten solder and to remove disruptive corrosion layers. This liquefies well below the actual solder temperature and supports the soldering process. In electronics, rosin with various additives is often used.

Leaded solders in electrical engineering

In the past, tin solders with different lead contents were industrially processed. The advantage over pure tin is the lower melting point of the alloys. In addition, tin solders have good mechanical toughness and formability, even at low temperatures. Since lead is a harmful heavy metal, lead solders are only used in the electrical industry in exceptional cases, and are generally replaced by lead-free tin solders.

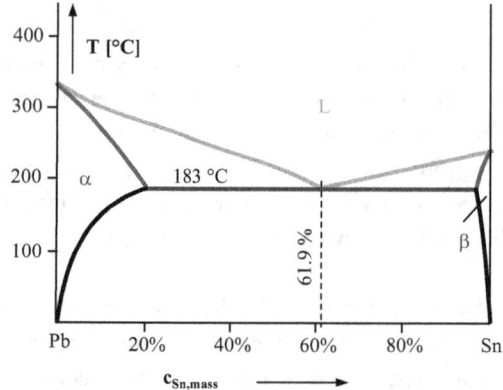

Alloy	Melting Range
Sn60Pb40	183 °C to 190 °C
Pb93Sn5Ag1.5	304 °C to 310 °C
Sn62Pb36Ag2	179 °C

Tab. 2.7: *Typical alloys used for soldering and their melting ranges*

Fig. 2.70: *Lead-Tin phase diagram*

Lead-free solders in electrical engineering

Since 2006, lead-free solders have mainly been used for soldering electronic assemblies. The EU's "RoHS" (Restriction of Hazardous Substances) directive, which was introduced for this purpose, has since regulated the use of certain hazardous substances in electrical and electronic equipment.

RoHS-compliant products may not contain more than 0.1 percent by weight of lead, mercury, hexavalent chromium or more than 0.01 percent by weight of cadmium per homogeneous material. When RoHS was first introduced, a number of exceptions were made for some industrial sectors (automotive industry, military and medical technology) that require high-reliability solders containing lead. However, these exemptions have now largely expired or are only applicable for a limited period of time. Similar regulations are also in preparation or already in force in countries outside the EU.

Tin-silver-copper alloys (lead-free solders) are used as a substitute for solders containing lead. They have the advantage that they do not contain lead, which is hazardous to health. However, their process properties are inferior to those of leaded solders. One of the main disadvantages of lead-free solders is their higher melting temperature. Lead-free solders have a melting temperature of 217 - 227 °C, which is approx. 40 °C higher than the melting temperature of solders containing lead (183 °C).

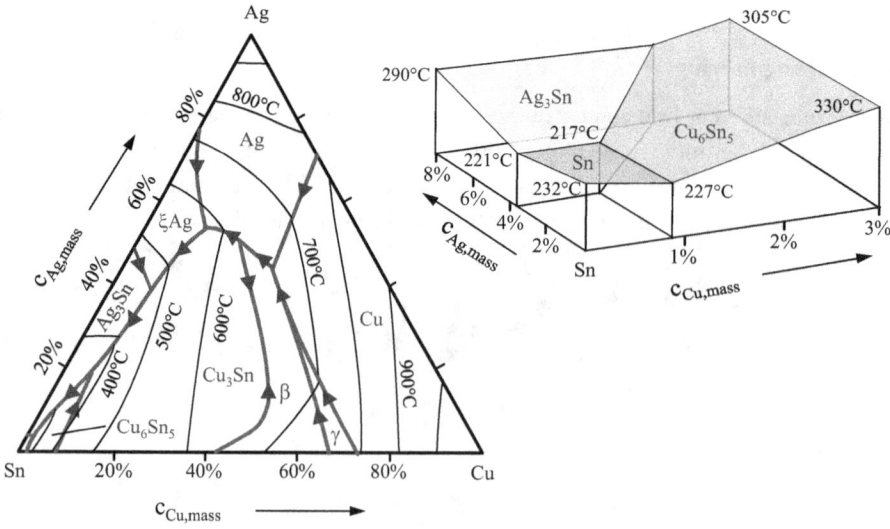

Fig. 2.71: *Phase diagram of tin/copper/silver-alloys (Liquidus temperatures)*

Fig. 2.71 shows the phase diagram of tin-copper-silver alloys and the section of the phase diagram used for lead-free soldering. The eutectic point of the alloy is at 217 °C for a silver content of about 4 % and a copper content of about 1 %. In industrial applications, the silver content is kept as low as possible for cost reasons. It can be reduced to 1 % without significantly increasing the melting temperature (see Tab. 2.8).

Alloy	Melting Range	Possible Additives
SnAg3Cu0.5	217 °C to 219 °C	Ni – improved solidification behavior, reduces copper separation, improve the flow properties, results in shiny solder joints
SnCu0.7Ni	227 °C	
SnCu0.7Ag1.0NiGe	217 °C to 219 °C	Ge – Antioxidant

Tab. 2.8: Alloys for lead-free soldering

There is a large variety of possible additives in lead-free solders. Very often, a small amount of nickel is found in the alloy, which improves the solidification behavior, reduces copper delamination, improves the flow properties, and produces shiny solder joints. The latter is helpful for optical quality control, since poorly soldered joints (which occurs when the solder was too cold during the soldering process and therefore did not melt completely) tend to give a dull optical appearance. Furthermore, germanium is added to the solder as an antioxidant.

Fig. 2.72 shows a typical solder profile for a reflow process with lead-free solder. In the heating phase, all components, the PCB and the solder are brought to a uniform temperature. At the same time, the flux is activated, which ensures the removal of thin oxide layers and good wetting of the molten solder on the copper surface. To keep the temperature load on the electronic assemblies as low as possible, the temperature profile exceeds the melting point of the solder only briefly.

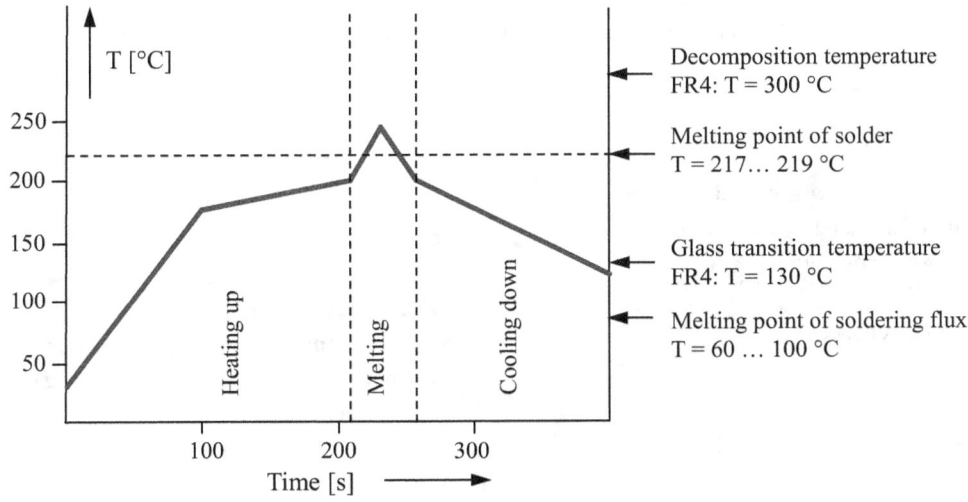

Fig. 2.72: Typical temperature profile of a reflow process with lead-free solder

3 Sensors

Sensors convert the physical parameter, which should be measured into an electrical signal. Fig. 3.1 shows the basic structure of sensors. In the first step, the physical parameter is detected and converted into an electrical signal. Then signal adjustment (filtering, amplification, error compensation) and conditioning of the output signal is performed.

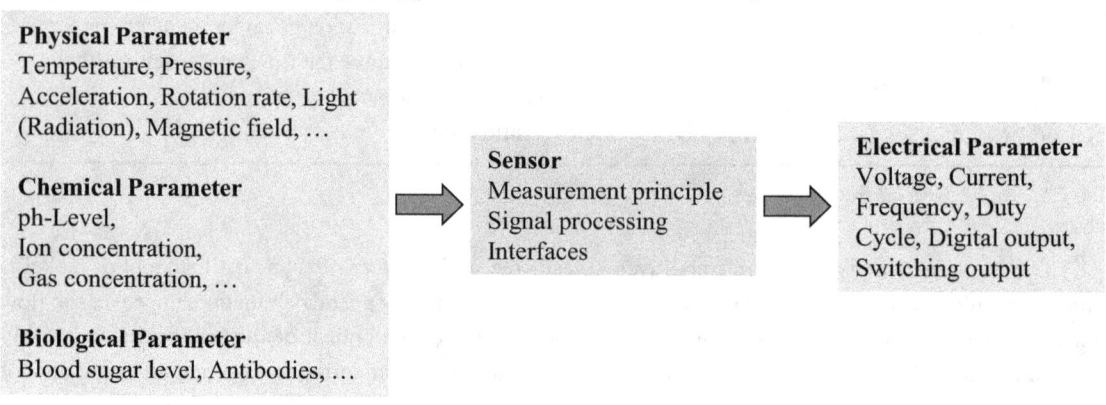

Fig. 3.1: *Sensor converting principle*

The measuring act is the most important process of this signal chain. It should have a very high sensitivity to the parameters to be measured, and a low cross-sensitivity to all other parameters. A difficulty for sensors is the necessary contact of the respective environment to the sensor chip. For electronic circuits, it is only necessary to couple in and out electrical signals, for sensors however, there also has to be the contact to the measured medium.

Sensors are the main field of microsystems technology. It enables the integration of electronics and sensor technology in smallest space. The following development trends for microsensors can be observed in recent years:

<u>Performance</u>
- **High accuracy without calibration**
- High linearity and dynamic range
- Low cross sensitivities
- High reliability
- Low maintenance

<u>Integration</u>
- **Digital Interfaces**
- Standardization of sensor interfaces
- Network ready
- **Multisensing**
- **Wireless output signals**
- Autonomous sensors / Energy harvesting
  - Sleep mode
  - Power management

<u>Smart Sensors</u>
- **Self-diagnosis**
- Self-calibration
- Self-identification
- Integrated signal processor

<u>Miniaturization</u>
- **Small size**
- **Low power requirement**
  - Low operating voltage
  - Low energy consumption

3.1 Acceleration Sensors

Acceleration sensors belong to the class of inertial sensors. They measure the effective acceleration acting on the sensor. Thereby, the origin of the acceleration is not of importance. In total, three acceleration types can overlap, which add vectorially.

- Changing of speed (a_1) $\qquad \vec{a_1} = \dfrac{dv}{dt}$ [7]

- Gravitation (a_2) $\qquad \vec{a_2} = 9.81 \text{ m/s}^2$ [8]

- Centrifugal acceleration (a_3) $\qquad \vec{a_3} = \dfrac{v^2}{r}$ [9]

Measured overall acceleration $\qquad \vec{a} = \sum_{i=1}^{3} \vec{a_i}$ [10]

The use of accelerometers cover a wide spectrum, because their use is unproblematic. Unlike pressure sensors, for example, accelerometers require no direct contact to the outside world. This allows their integration in normal IC packages. In addition, it is beneficial that, in the case of rigid, linearly moving bodies, the acceleration is the same at every point on the body, thus enabling the flexible positioning of the sensor in the measurement object.

One of the main applications of acceleration sensors is the automotive industry. Already in the 1990s, airbag sensors were used, which gave the signal for firing the airbag in the event of an accident. Modern motor vehicles have up to 10 airbags integrated. The number of built-in acceleration sensors is therefore of the same magnitude. More operation sites of acceleration sensors in the motor vehicle are, for example, in the adaptive chassis control or alarm systems. Considering that Volkswagen alone produces about 5 million cars every year, the number of acceleration sensors in this field is over 1 billion sensors worldwide. With these kinds of numbers, it is possible, in spite of enormous development costs, to produce sensors that are in the one-dollar range. Other industries, such as the consumer market or the medical industry, in which development would not be worthwhile because of lower volumes, profit here from the developments in the automotive industry.

Largely unnoticed, but not less interesting from the production quantities, is the use in the IT sector. Here acceleration sensors are used as crash sensors in hard drives, to move the heads on a landing zone in case of a shock event. Therefore a free-fall detection is used, were the absence of the gravitational force during free-fall is detected.

The largest sales market for accelerometers are smartphones. In the beginning, there were only simple functions carried out, for example the recognition of shaking, or the detection of the orientation of the display. But currently, mostly in combination with gyroscopes and magnetic sensors, more complex motion detections are possible. Through this field of application the development pressure towards even smaller and very flat sensor packages has grown enormously.

Piezoelectric Acceleration Sensor

The classic among the acceleration sensors is the piezoelectric sensor type, which is produced by precision manufacturing techniques. Fig. 3.2 shows the structure of such a sensor. For the piezoelectric material, usually Lead-Zirconate-Titanate (PZT) ceramics are used. Since the net mass of the very thin ceramic plate is very low, often an additional mass is mounted to increase the sensitivity. This causes a much higher force on the piezoelectric disk at the same external acceleration. Thus, the generated voltage, and the sensitivity, is increased. On the other hand, the upper limit frequency of the sensor is reduced by the additional mass, which is not always desirable.

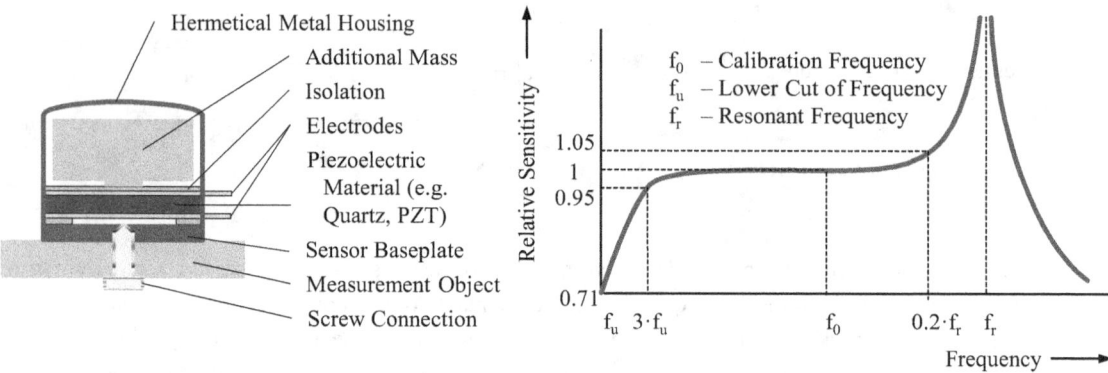

Fig. 3.2: Structure and typical frequency response of a piezoelectric accelerometer

Measurement range	500 g
Sensitivity	10 mV/g
Noise level	0.8 mg
Resonance frequency	32 kHz
Frequency range	1 to 10000 Hz
Temperature range	- 51 °C to 100 °C
Weight	8.7 g
Dimensions	15 x 12 x 19 mm^3

Fig. 3.3: Use of a piezoelectric accelerometer as a reference sensor on the vibration test bench

Tab. 3.1: Key data of the piezoelectric acceleration sensor Deltron 4513 by Brüel & Kjær

Structure of MEMS Accelerometer

MEMS accelerometers are based on spring-mass systems and are usually read out by a capacitive method. Both the spring-mass system and the capacitive readout structures are integrated into a silicon functional layer. Fig. 3.4 shows the process of the conversion within the sensor from the acceleration (a) to the output voltage signal (ΔV). Hereinafter will be described the individual steps of this conversion.

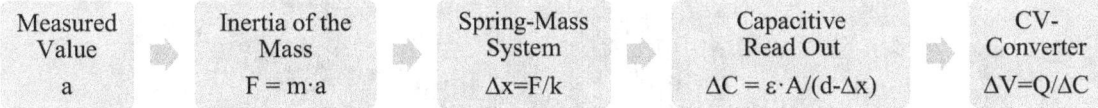

Fig. 3.4: Conversion chain of an acceleration sensor

Fig. 3.5 shows the basic structure of a surface micromechanical 1-axis acceleration sensor. Shown in red are the springs, which allow a deflection of the mass only in the x-direction. The springs usually have thicknesses of 1 to 2 microns and are made of polycrystalline silicon. Within the same layer are also the mass and the capacitive evaluation structures. The fixed points of the springs represent the anchor structures, which have been grown on the silicon substrate. If acceleration in the x-direction occurs, an inertial force acts on the mass (Eq. [13]) and deflects the spring-mass structure opposite to the direction of the acceleration (Eq. [17]). Ensuring that acceleration overload does not damage the very precise capacitive readout structures, limiters are installed in the sensor structure. They should also prevent fingers from the readout structure from coming in contact and sticking together (sticking effect).

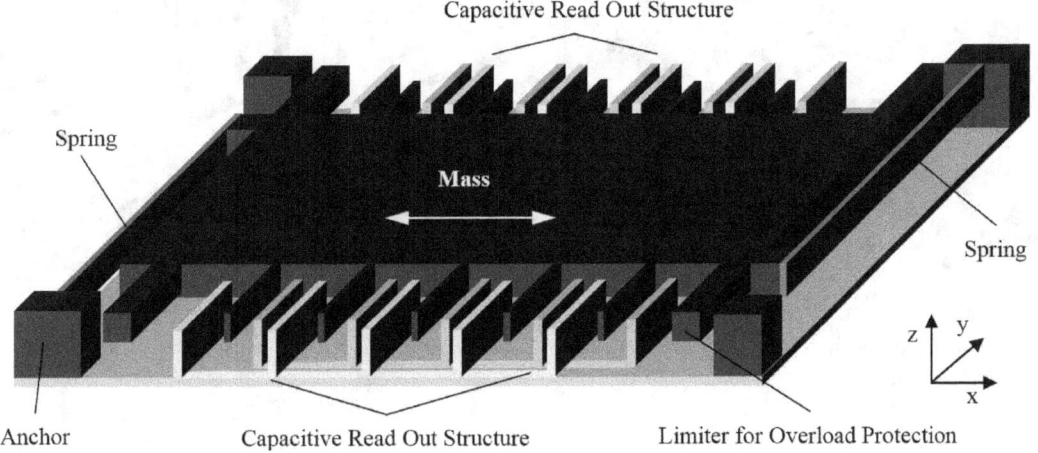

Fig. 3.5: Schematic representation of a micromechanical acceleration sensor with capacitive readout structure

Fig. 3.6 shows the polysilicon structure of the 3-axis accelerometer ADXL337, which consists of a large-area mass suspended from four polygon springs. The x- and y-deflection of the mass is read out by two capacitive evaluation structures each, which operate according to the differential capacitor principle. Furthermore, the sensor contains a flat electrode below the mass structure, which aids the measurement of the z-deflection.

The polysilicon structures are 7.5 µm thick and have a horizontal structure resolution of 1.5 µm.

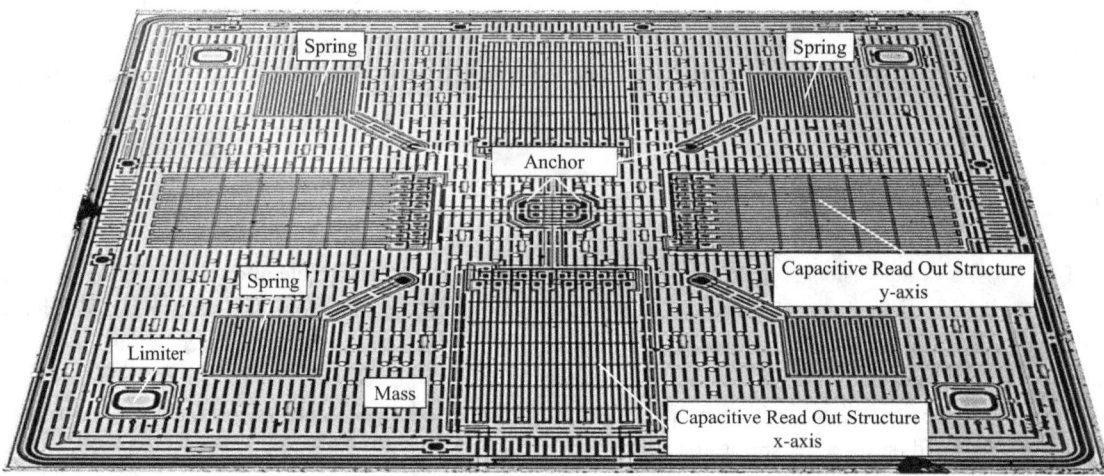

Fig. 3.6: 3-axis accelerometer ADXL337 from Analog Devices

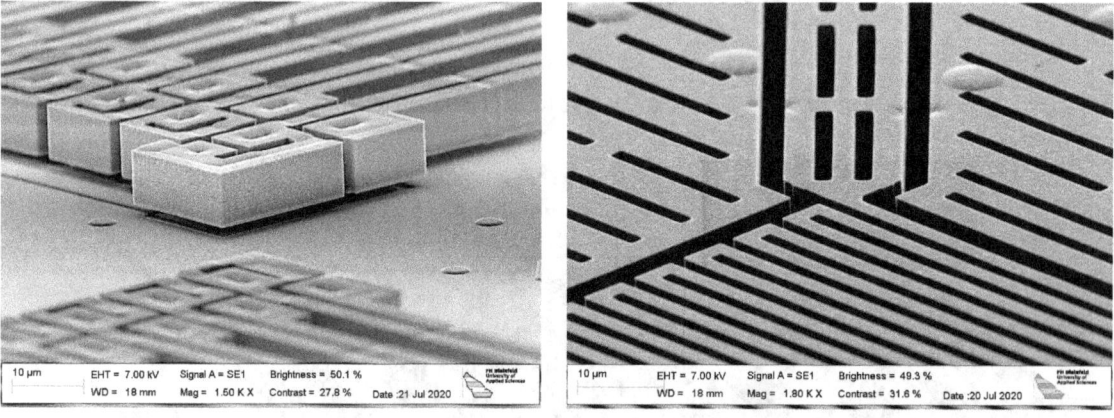

Fig. 3.7: Detailed view of the capacitive evaluation structure (left) and the spring structure (right) of the ADXL337 accelerometer.

Static and Dynamic Behavior Model of a Spring-Mass System

To better understand the real behavior of an acceleration sensor, it is helpful to take a closer look at the abstract spring-mass model. Fig. 3.8 shows this model. It consists of a spring, a mass and a damping element. In this simplified model, the mass has no spatial extent (point mass) and is defined only by its weight (m). The spring is completely linear and will be described by Eq. [11]. Furthermore, the damping element in the first approximation is independent from the displacement speed Eq. [12].

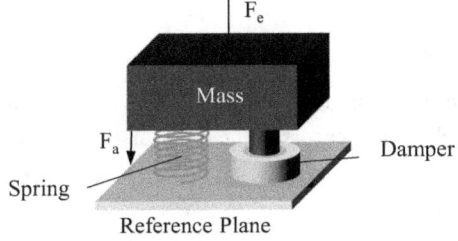

Fig. 3.8: Simplified model of a spring-mass system

$F = k \cdot x$ [11]

$F = c \cdot v$ [12]

with x - Deflection of the mass
      v - Velocity of the mass
      c - Attenuation constant
      k - Spring constant

Spring-mass systems can be mathematically described by a 2nd-order differential equation.

$F_e(t) = m \cdot a(t)$ [13]

$$m\frac{d^2x}{dt^2} + c\frac{dx}{dt} + kx = F_e(t)$$ [14]

Where F_e is the force on the mass, which is exerted by the acceleration to be measured and $\frac{d^2x}{dt^2}$, $\frac{dx}{dt}$, x describe the movement and the location of the mass. The dimensions of the spring and the young-modulus of the spring material have a significant influence on the spring constant. Eq. [15] exemplifies the analytical formula to calculate the spring constant of a cantilever [5].

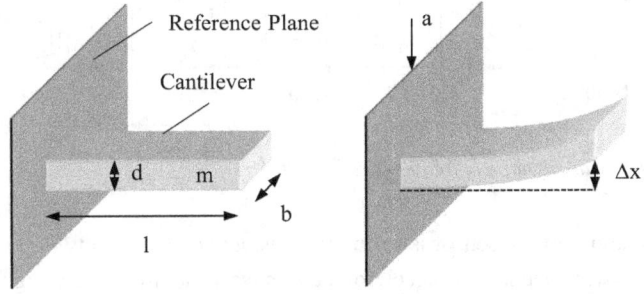

Fig. 3.9: Deflection of a cantilever

---

[5] U. Mescheder, Mikrosystemtechnik, Teubner Verlag, 2004

$$k = \frac{E \cdot b \cdot d^3}{4 \cdot l^3} \qquad [15]$$

$$\Delta x = \frac{4a \cdot m \cdot l^3}{E \cdot b \cdot d^3} \qquad [16]$$

with E - Young-module
      m - Mass
      k - Spring constant

For static accelerations, Eq. [11] can be simplified and it is obtained:

$$x_{statisch} = \frac{m \cdot a}{k} \qquad [17]$$

It is seen that the deflection of the mass is proportional to the acceleration to be measured. This applies until the maximum possible deflection of the mass, which lies for acceleration sensors in the order of 1 to 10 µm.

Following is shown the solution of the equation in the frequency domain:

$$H(s) = \frac{x(s)}{a(s)} = \frac{1}{s^2 + \frac{c}{m}s + \frac{k}{m}} = \frac{1}{s^2 + \frac{\omega_0}{Q}s + \omega_0^2} \qquad [18]$$

$$\omega_0 = \sqrt{\frac{k}{m}} \qquad Q = \frac{\sqrt{k \cdot m}}{c} \qquad \omega_R = \omega_0\sqrt{1 - \left(\frac{c}{2m \cdot \omega_0}\right)^2} \qquad [19]$$

with H   - Complex transfer function
      s   - Complex frequency variable (jω)
      Q   - Quality
      ω   - Angular frequency (2πf)
      ω_0 - Eigenfrequency
      ω_R - Resonant frequency

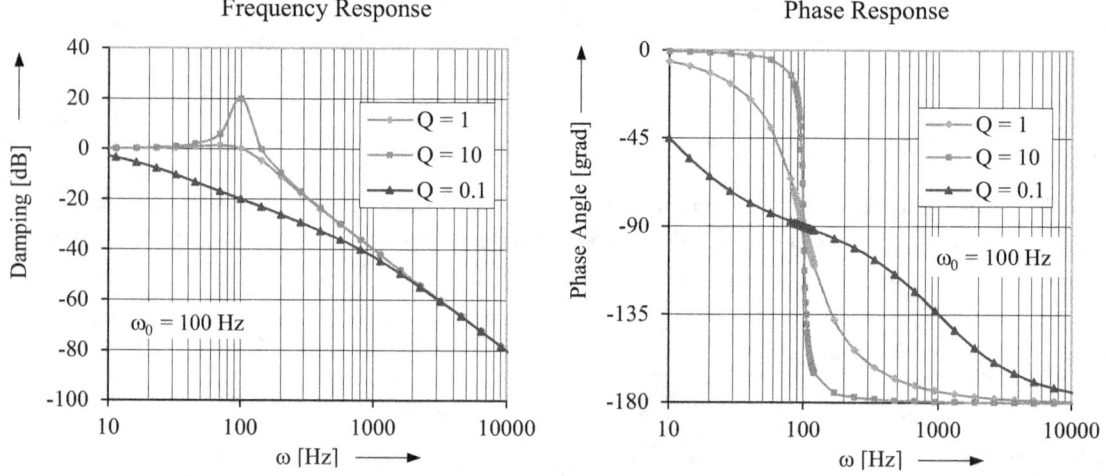

Fig. 3.10: Dynamic behavior of a spring-mass system with the natural frequency $\omega_0 = 100$ Hz

As can be seen in Fig. 3.10, the dynamic behavior of the sensor is strongly dependent on the quality of the system and therefore of the damping constant. Micromechanical accelerometers have generally a very high damping and a quality Q < 1. For attenuation especially, air damping within the capacitive readout structures is responsible. Since these have only a very small gap (1 to 2 µm) and at the same time a relatively large area, the air cannot escape fast enough from the gap, and is compressed during a movement. This means that power is withdrawn from the spring-mass system and so the oscillation thereof is damped. Another damping would be the structural damping within the silicon spring. It is, however, in relation to the air damping negligible.

An Analogy of Mechanical to Electrical Systems

Electrical engineers find it difficult to understand mechanical systems. Therefore, it is common to represent the mechanical behavior by analogy considerations with electronic systems. In this case, the behavior of the damped spring-mass system is analogous to the lossy LC-low-pass filter. The needed transformation laws for it are listed in Tab. 3.2.

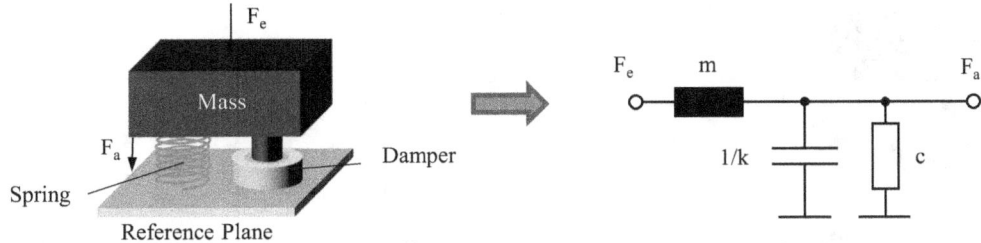

Fig. 3.11: Simplified model of a spring-mass system and LCR-low pass as the counterpart from electrical engineering

		Force	Velocity	Spring	Mass	Damper
Mechanically		F [N]	v [m/s]	k [kg·s^{-2}]	m [kg]	c [kg·s^{-1}][N·s/m]
				$F = k \int v \, dt$	$F = m \dfrac{dv}{dt}$	$F = c \cdot v$
		Voltage	Current	Capacitance	Inductance	Resistivity
Electrically		V [V]	I [A]	C [As/V] [F]	L [V/As] [H]	R [V/A] [Ω]
				$U = \dfrac{1}{C} \int I \, dt$	$U = L \dfrac{dI}{dt}$	$U = R \cdot I$

Tab. 3.2: An analogy between mechanical and electrical variables

Capacitive Readout Principle

MEMS accelerometers very often have capacitive readout structures, in which the deflection of the mass is detected by a change in capacitance. Fig. 3.12 shows a simple single capacitor structure with an initial capacitance C_0. If the mass and the coupled movable electrodes are deflected by an external acceleration for a distance of Δx, there is a change in the capacitance of ΔC.

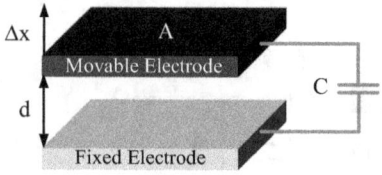

Fig. 3.12: Singe capacitor structure

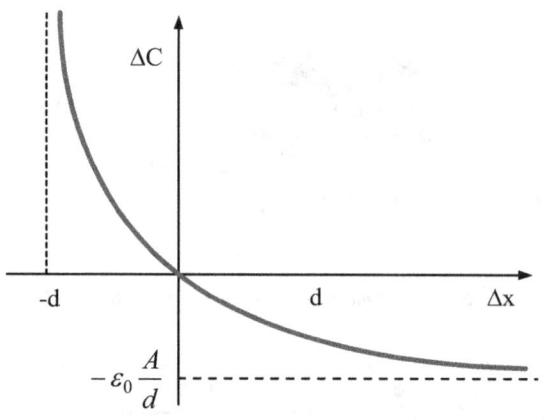

Fig. 3.13: Characteristic of the single capacitor

$$C_0 = \varepsilon_0 \frac{A}{d} \qquad [20]$$

$$\Delta C = A\varepsilon_0 \frac{-\Delta x}{d(d + \Delta x)} \qquad [21]$$

$$\Delta C = A\varepsilon_0 \frac{-\Delta x}{d^2} \quad \text{für } \Delta x \ll d \qquad [22]$$

For small deflections ($\Delta x \ll d$) the change in capacitance is proportional to the deflection and thus well suited for a linear sensor characteristic. However, larger deflections elicit non-linearities and asymmetries between positive and negative displacements, and thus a different behavior of the sensor for positive and negative acceleration values.

For this reason, differential capacitor structures are usually used in today's accelerometers. Fig. 3.14 shows such a structure. The individual capacitances (C_1, C_2) in this arrangement are not evaluated, but the differential capacitance (C_{Diff}). Considering the transfer curve in Fig. 3.15, it is seen that with this arrangement, the problem of asymmetry has been solved. However, the non-linearity of the characteristic curve remains at large deflections, but the range in which the linear behavior is seen is significantly increased compared to the single capacitor structure.

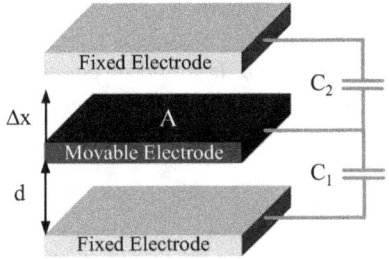

Fig. 3.14: Differential capacitor structure

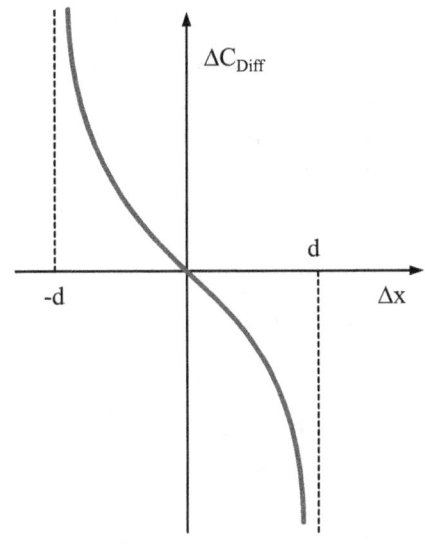

Fig. 3.15: Characteristic of a differential capacitor structure

$$C_{Diff} = C_1 - C_2 \qquad [23]$$

$$C_{Diff} = A\varepsilon_0 \frac{-2\Delta x}{d^2 - \Delta x^2} \qquad [24]$$

$$C_{Diff} = A\varepsilon_0 \frac{-2\Delta x}{d^2} \quad \text{für} \ \Delta x^2 \ll d^2 \qquad [25]$$

The achievable capacitance of a single capacitor arrangement in the surface micromachining is relatively low (fF-range). For this reason, several structures (10 to 100) are connected in parallel, so a total value in the range of pF can be reached. Fig. 3.16 shows such capacitive readout structures (comb structures) of the sensor ADXL202.

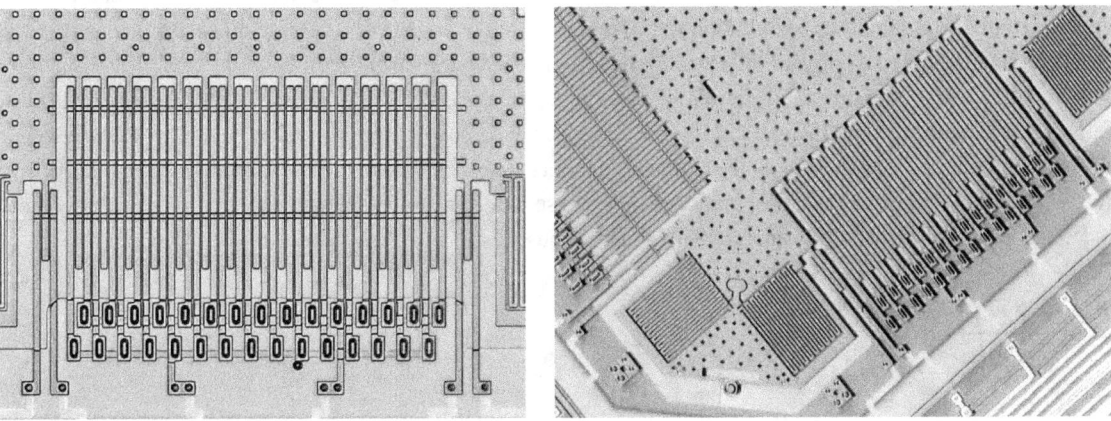

Fig. 3.16: Capacitive readout structures of the acceleration sensor ADXL202 from Analog Devices

Capacitance-to-Voltage Converter (CV-Converter)

There are various methods to measure the capacitance of an acceleration sensor. It would be possible to measure the impedance of the sensor, to determine the discharge constant, or to build, with the help of an inductor, a resonant circuit and measure the resonant frequency. However, for capacitance measurements in the fF-range, the charge pump principle is more suitable.

Fig. 3.17 shows such an arrangement to evaluate a differential capacitor structure.

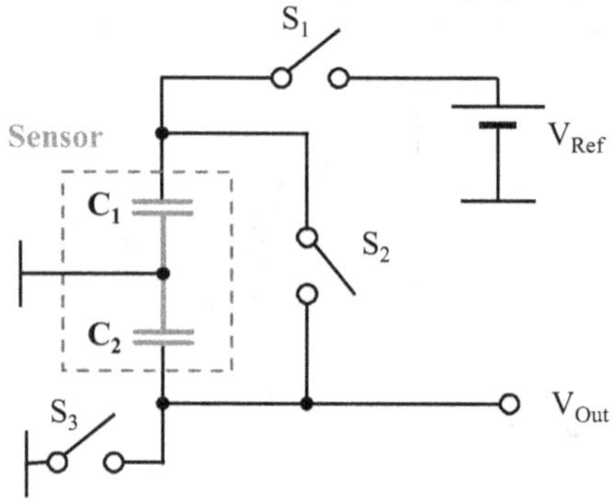

Step	S1	S2	S3		
1	x	o	o	$Q_1 = C_1 \cdot U_{REF}$	[26]
2	o	x	o	$U_A = \dfrac{C_1}{C_1 + C_2} U_{REF}$ $U_A = \left(1 + \dfrac{\Delta C}{C_0}\right) \dfrac{U_{REF}}{2}$ with $C_1 = C_0 + \Delta C$ $C_2 = C_0 - \Delta C$	[27] [28]
3	o	o	x	$Q_2 = 0$	[29]

Fig. 3.17: Circuitry and switching sequence of a charge pump for measuring the capacitance difference

The operation is shown in the adjacent table. The measurement of ΔC is done in three steps. In Step 1, C_1 is charged on V_{REF} and so a charge Q_1, according to Eq. [26], is stored in C_1. In Step 2, S_1 is opened and S_2 is closed. C_1 and C_2 are now connected in parallel, and the charge Q_1 is distributed over both capacitors. The resulting capacitor voltage is, according to Eq. [28], proportional to the change in the sensor capacitance. The capacitor voltage is sampled at the output (V_{Out}) and is stored by means of a sample-and-hold device. In Step 3, C_2 can be discharged again and the measurement sequence restarts from Step 1.

Another way to measure the differential capacitance of the acceleration sensor is shown in Fig. 3.18. In this case, a measuring signal is applied to both capacitor plates of the sensor, which is 180°-phase-shifted. If the capacitors have the same size, both signal components are canceled out at the center electrode of the sensor. In the case of $C_1 > C_2$, there is a positive, or in the reverse case, a negative signal generated. The voltage signal is tapped at the center electrode and scanned synchronously.

$$U_A = \left(\frac{2 \cdot C_1}{C_1 + C_2} - 1\right)\hat{U}_G = \frac{\Delta C}{C_0}\hat{U}_G \qquad [30]$$

with $\hat{V}_G$ - Generator voltage (peak value)
C_0 - Initial capacitance of C_1 and C_2

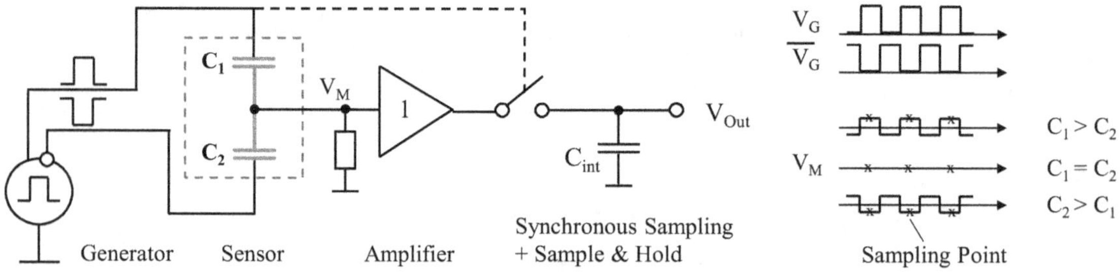

Fig. 3.18: CV-converter to readout differential capacities

A very special measurement method for determining the differential capacitance of acceleration sensors is the closed loop principle. Here, a voltage is applied to the external electrodes, which keeps the mass in the middle by the resulting electrostatic attraction forces. At the beginning of a cycle, it is determined, whether the mass is deflected by an extern acceleration, during a measuring phase. If this is the case, an electrostatic force is initiated, which brings the mass back to the center position. At the end a capacitive evaluation system is formed, in which the sensor is always operated in or near the zero position, so nonlinearities can be minimized. Another positive aspect is the acquisition of the spring function by the electrostatic control system. Thereby the technological variations of the spring dimensions have only a minor influence on the sensor characteristics.

A disadvantage of this evaluation is that a relatively high voltage is needed to generate the return force. This is also the reason why this type of sensor evaluation is rarely used nowadays. Today's sensors operate with a supply voltage down to 1.8 V, so providing voltages to generate a sufficient electrostatic force is very difficult. In addition, the power consumption of a closed-loop circuit is significantly higher than for passive evaluation methods.

$$F = \frac{1}{2} \frac{\varepsilon_o \cdot A}{d_0^2} \cdot (U_{C2}^2 - U_{C1}^2) \qquad [31] \qquad \text{with } F \text{ - Electrostatic force}$$

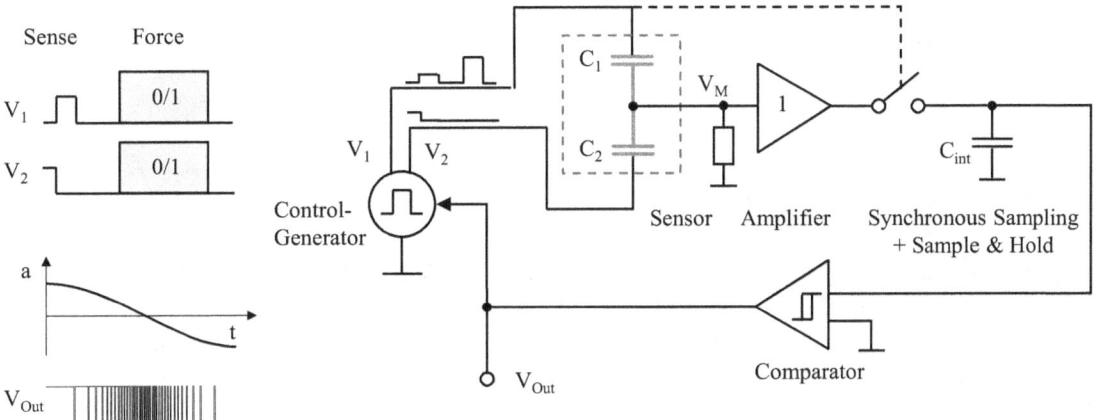

Fig. 3.19: Schematic view of a closed-loop CV-converter

Fig. 3.19 shows the structure of a closed-loop CV converter. The sense phase of the measurement procedure corresponds to the passive converter, shown in Fig. 3.18. It determines whether the mass was deflected by an acceleration. The second phase (force-phase) is new, where a voltage is applied on one capacitor plate depending on the results of the sense-phase. The force phase aims to return the sensor mass to its center position by the applied electrostatic attraction force.

The output signal of the CV converter is the drive signal of the position control. It represents a pulse width modulated signal and is proportional to the acceleration to be measured.

Parameters of Acceleration Sensors

In the following chapter, the main parameters of an acceleration sensor are presented. In addition to the metrological descriptions of the individual parameters, the parameters of the sensor ADXL 203[6] are shown to illustrate the magnitudes of the parameters.

Sensitivity

The ideal acceleration sensor has a linear dependence between the output voltage and the acceleration. This dependence is described by the sensitivity (S). The maximum output voltage is obtained at the full-scale acceleration value (FS) with S·FS.

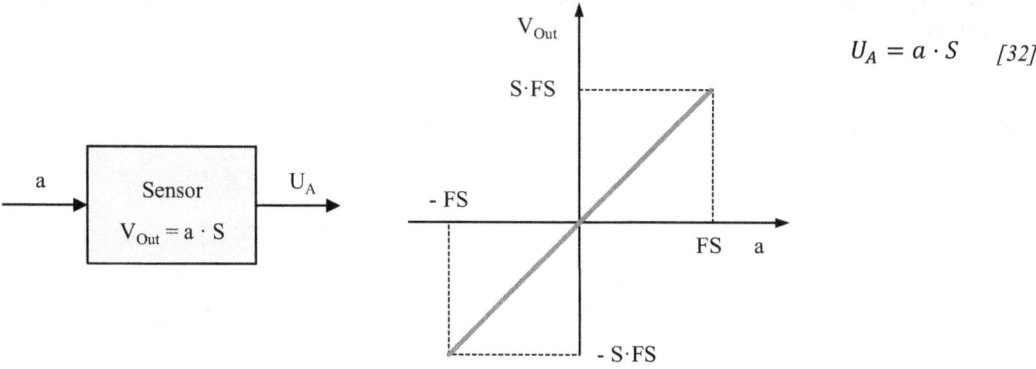

$$U_A = a \cdot S \quad [32]$$

Fig. 3.20: Transfer characteristic of an ideal acceleration sensor

For non-calibrated sensors, the dispersion of sensitivity is relatively high. This arises mainly due to the technological variations of the spring thickness and the gap spacing of the comb structures. For this, the manufacturer indicates mostly the three values (S_{min}, S_{typ} and S_{max}) in the data sheet. For the example sensor ADXL203, one finds a scattering of $S_{min} = 960$ mV/g up to $S_{max} = 1040$ mV/g with a typical sensitivity of $S_{typ} = 1000$ mV/g.

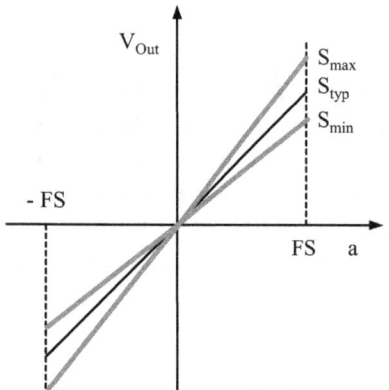

Fig. 3.21: Production-related scattering of sensitivities

---

[6] Data sheet ADXL203, Analog Devices, 2004

In addition to the limits of the sensitivity (S_{min} and S_{max}), the manufacturer often describes the scattering by the standard deviation (σ). Prerequisite for this is that the sensitivity is Gaussian-shaped spread. In Fig. 3.22, the distribution of the sensitivities is shown for 100 test samples of the ADXL203. Almost all samples are in the 3σ area. Theoretically, this should be 99.7 % of all expected sensitivities. If an error rate of 0.3 % is tolerable by the user, the possible variation range of the sensitivity can be significantly limited, compared to the S_{Min}-S_{Max}-area.

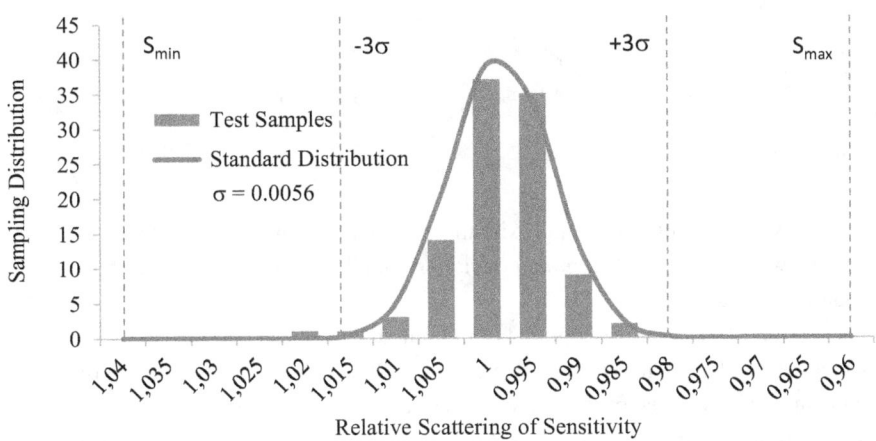

Fig. 3.22: Representation of the scattering of the sensitivity for 100 random samples of the ADXL203

A special feature results for the ADXL203 through its ratiometric output. In this type of output, the sensitivity is directly proportional to the supply voltage. The values in the data sheet of the ADXL203 are specified for an operating voltage of 5 V. For other operating voltages, the sensitivity is specified according to Eq. [33].

$$S = \frac{V_S}{5\,V} S_{5V} \quad [33]$$

with S_{5V} - Sensitivity for a supply voltage of 5 V
V_S - Supply voltage

In addition to the production-related scattering of the value, there is also a temperature dependence for the sensitivity. It is described by the temperature coefficient, or by the maximum percentage change. A maximum variation of ± 0.3 % is reported over the entire temperature range (40 °C ≤ T ≤ + 125 °C) as typical value for the ADXL203.

The previous considerations assume that the transfer curve is a straight line. In practice, however, the values spread around the ideal line, and characteristic curvatures occur, especially at the end of the measuring range.

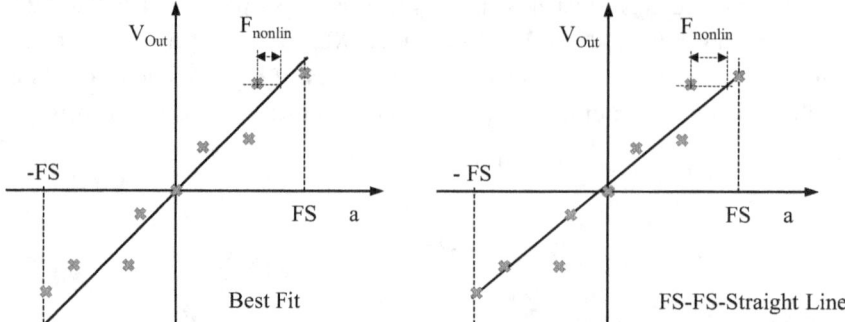

Fig. 3.23: Nonlinearity of the sensor characteristic

Fig. 3.23 shows two ways of a straight-line approximation for scattered measurement points. The best-fit method determines the line with the smallest summary deviation for all measuring points. In this case, usually the method of least squares is used.

Another method is to connect the measuring points of the full-scale values by a straight line. It is easier to carry out, but results in a slightly larger nonlinearity value (F_{nonlin}). For this reason, usually the first method is chosen by the sensor manufacturers to determine the non-linearity of the data sheet. For a maximum value of $F_{nonlin} = \pm 1.25$ % FS according the data sheet of the ADXL203, a maximum error of ± 0.02 g occurs.

Frequency Response

As described in the previous section, spring-mass systems have a second order low-pass behavior, resulting in frequency dependence of the sensitivity. Furthermore, an adjustable low-pass filter 1st order is also integrated in most accelerometers. With them, it is possible to adjust the user-defined frequency response of the acceleration sensor, independently from the resonant frequency of the spring-mass system. Fig. 3.24 shows an example of the frequency response of such a sensor. It can be seen in the theoretical course on the left side, that up to the cut-off frequency, the sensitivity corresponds to of the static sensitivity. Then it falls by - 20 dB per decade up to the resonant frequency. From here, the filter behavior of the electronic low-pass filter and the spring-mass system add up, and results in a drop of - 60 dB per decade. Depending on the quality of the spring-mass system, it is possible that resonant peaks occur at this frequency. The example shown in Fig. 3.24 describes a system with a cut-off frequency of 500 Hz, a resonance frequency of 5.5 kHz and a critical damping (Q = 0.5). On the right side of the figure the measured frequency response of a sensor board (PCB 10 x 10 cm²) is presented, which has the same frequency settings. Note the individual resonant peaks of the circuit board, which almost completely covers the theoretical characteristic curve of the sensor. The sensor board sample illustrated here is mounted at only two points (see Fig. 6.11) to highlight the resonant peaks of the PCB. The example demonstrate clearly, that a rigid attachment is necessary for acceleration measurements under vibration influence.

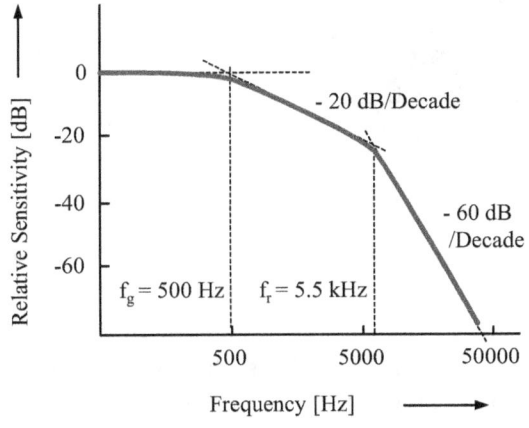 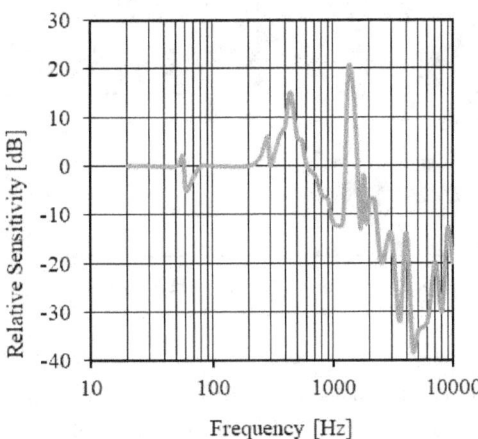

Fig. 3.24: Theoretical frequency response of the sensitivity of an acceleration sensor with a resonance frequency (f_r) of 5.5 kHz and a set cut-off frequency (f_g) of 500 Hz and the measured frequency response of a sensor board having the same frequency parameters

Noise

A quality characteristic of acceleration sensors is the lowest possible noise. The lower the noise floor of the sensor, the smaller the acceleration that can be detected. The noise is caused mainly by the electronics, which must determine capacitance changes in the fF-area. In addition, a noise component arises from the Brownian motion of gas molecules in the capacitive evaluation structures.

For a description of the output noise, the noise density [µg/√Hz] is often indicated. It describes a white noise with a frequency-independent noise density. For the ADXL203 a value of S_R = 110 µg/√Hz can be found in the data sheet. For a given bandwidth, the resulting output noise is obtained after Eq. [34], whereby the bandwidth is defined by a 1st-order low-pass filter. The smaller the cut-off frequency of the low-pass filter, the smaller the output noise. In addition to the white noise other noise components occur, which cannot be described with the noise density. An example is the 1/f-noise. So a minimal noise output voltage is specified in the datasheet independently from the bandwidth. For the ADXL203 this is u_R = 1 mV.

$$V_R = S_R \cdot \sqrt{f_g \cdot 1{,}6} \qquad [34]$$

Equation [34] shows the ratio of the noise output voltage (u_R) to the noise density. The factor of 1.6 describes the non-abrupt truncation of the frequency band above f_g by the low-pass of the 1st order. It should be noted that u_R is an rms-value. Especially for acceleration measurement, applications occur where often a maximum acceleration limit should be monitored. For such applications, it must be considered that the peak value of the noise voltage can be significantly larger than the calculated effective value (Tab. 3.3).

$V_{Out, Peak\ to\ Peak}$	Percentage of time that noise exceeds nominal peak-to-peak value
$2 \cdot v_R$	32 %
$4 \cdot v_R$	4.6 %
$6 \cdot v_R$	0.27 %
$8 \cdot v_R$	0.006 %

Tab. 3.3: *Estimation of Peak-to-Peak Noise*

Fig. 3.25 shows an example for noise of the acceleration sensor ADXL202. For the bandwidth of 500 Hz, a theoretical noise voltage of $v_R = 1.76$ mV results for this sensor. Despite the still superimposed ripple voltage with a ripple around 2 mV, it can be easily seen in this graph, that individual values exceed 4 times the rms-value.

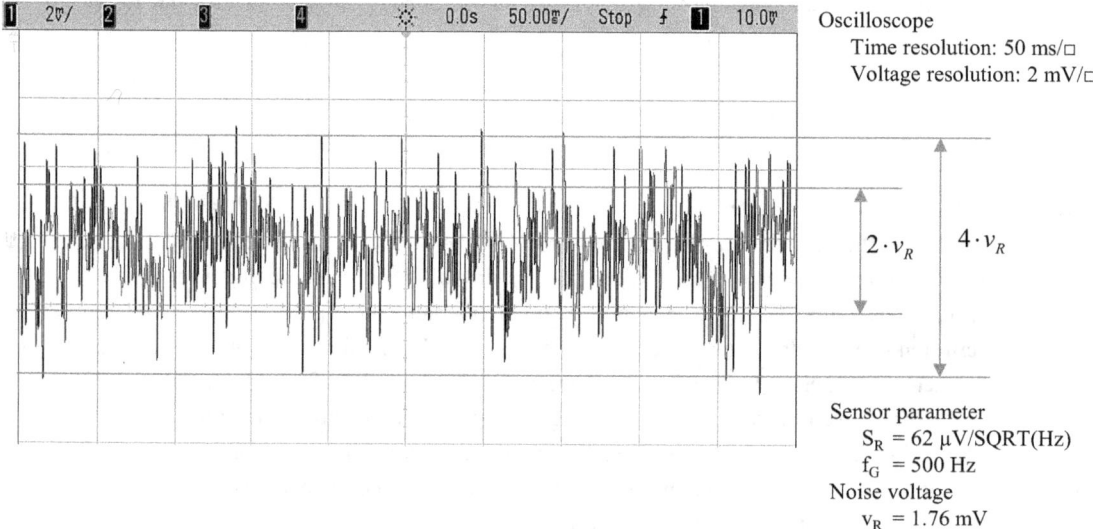

Fig. 3.25: *Noise output voltage of the acceleration sensor ADXL202*

Cross-Axis Sensitivity

The cross-axis sensitivity of an acceleration sensor describes the sensitivity of the sensor to accelerations that occur perpendicular to the measurement axis. For the x-direction of measuring these are accelerations in y- and z-direction. Often the cross-axis sensitivity (S_{cross}) is expressed as the percentage ratio to the measurement sensitivity. For the ADXL203 a maximum cross-sensitivity of ± 3 % can be found in the data sheet.

$$S_{cross} = \frac{\Delta V_{Out}}{a_{cross}} \quad [35]$$

$$S_{cross,\%} = 100\% \frac{S_{cross}}{S} \quad [36]$$

with S_{cross} - Cross-Axis Sensitivity
$S_{cross,\%}$ - Relative percentage cross-sensitivity
ΔV_{Out} - Change of the output voltage by a_{cross}
a_{cross} - Acceleration perpendicular to the measuring axis
S - Sensitivity of the sensor along the measuring axis

Ideally, the cross-axis sensitivity is equal to zero. The undesired cross-axis sensitivity is primarily caused by tilting the measuring axes in relation to the sensor axis. A tilt of about 1 degree causes a cross-axis sensitivity of 1.7 %, for example. Bearing in mind that the orientation of a very small IC is very difficult to reproduce on the circuit board, for practical applications usually a larger cross-axis sensitivity of the overall system is obtained as specified in the sensor datasheet.

Zero g Bias Level

For static applications, such as the spirit level function, the stability of the zero point is the crucial parameter of a sensor. It is influenced by the temperature and the long-term drift of the sensor. For the former, a temperature dependence is specified by most manufacturers.

$$\Delta a_T = \alpha_T \cdot \Delta T \quad [37]$$

The sensor ADXL203 has a maximum value of $\alpha_T = \pm 0.8$ mg/°C. As a result, the maximum deviation of the zero point is ± 80 mg for the entire temperature range (- 40 °C ≤ T ≤ 125 °C). If this value is too large for an application, it is possible to compensate the temperature response with external electronics, mostly in form of a microcontroller. For this purpose, the temperature response of each sensor must be determined and then be compensated based on an internal temperature measurement. Information to the long-term drift of the zero point are not specified by most manufacturers.

Parameters of Various Acceleration Sensors

In Tab. 3.4 the parameters of three acceleration sensors from Analog Devices are listed exemplarily, which characterizes the development from 2004 to 2020. In addition to a tendency to improve the performance parameters (offset and sensitivity stability, noise density), there is a trend towards smaller designs and lower supply voltages. In addition, digital evaluations and interfaces with 16-bit resolution have established themselves in acceleration sensors.

	ADXL 203 Analog Devices	ADXL362 Analog Devices	BMA 456 Bosch
Measuring range	2-axis ± 1.7 g	3-axis ± 2 g, ± 4 g, ± 8 g	3-axis ± 2 g, ± 4 g, ± 8 g, ± 16 g
Sensitivity S	1000 mV/g (V_{DD} = 5V)	1 mg/LSB (2g-range)	0.06 mg/LSB (2 g-range)
Resolution		12 bit	16 bit
Non-Linearity F_{Nonlin}	0.2 % FS	± 0.5 % FS	± 0.5 % FS
Noise Density S_R	110 µg/$\sqrt{Hz}$	550 µg/$\sqrt{Hz}$ (x/y- axis) 920 µg/$\sqrt{Hz}$ (z- axis)	120 µg/$\sqrt{Hz}$
Cross-axes Sensitivity $S_{cross\%}$	± 1.5 %	± 1.5 %	0.5 % FS
Sensor Resonant Frequency	5.5 kHz	3.5 kHz	1.6 kHz (max. bandwidth)
Bias Offset Off	± 100 mg	± 150 mg (x/y-axis) ± 250 mg (z-axis)	± 20 mg
Bias Temperature Coefficient dOff/dT	0.1 mg/K	0.5 mg/K (x/y-axis) 0.6 mg/K (z-axis)	± 0,2 mg/K (x/y-axis) ± 0.35 mg/K (z-axis)
Interface	Analog	SPI	SPI, I²C
Supply Voltage	3.0 ... 6.0 V	1.6 V ... 3.5 V	1.6 V ... 3.6 V
Power Consumption	700 µA	1.8 µA for 50 Hz bandwidth	150 µA
Package	LCC 5 x 5 x 2 mm³	LGA 3 x 3 x 1 mm³	LGA 2 x 2 x 0.65 mm³
Release Year	2004	2012	2017

Tab. 3.4: Data comparison of different generations of acceleration sensors (typical values)

Sensors

Application of Accelerometers as Tilt Sensor

A common application of accelerometers is the measurement of the angle of an inclined plane. Either the resulting gravitational acceleration is measured and used to determine the angle of inclination or the angle is determined by the direction of the acceleration due to gravity field. Because the inclination measurement is a static measurement, the sensors used for this measurement should have the following characteristics:
- Low-g sensors (measuring range 1 to 2 g),
- High resolution,
- Excellent zero stability,
- Low cut-off frequency.

On the other hand, for measurements of the inclination in moving systems, such as on a moving pendulum, a combination of acceleration and gyroscopes have to be used.

Further applications for accelerometers are:
- Shock detection in the automotive and transport sector,
- Inertial navigation systems,
- Vibration measurements on machines,
- Motion monitoring of humans and animals,
- Active Loud speakers or
- Hard-disk crash sensors.

Tilt Measurement by Means of a Single-Axis Accelerometer

For inclination measurement with a single-axis acceleration sensor, the horizontal acceleration is measured and considered in relation to the acceleration due to gravity. This gives the sinusoidal dependency shown in Fig. 3.27.

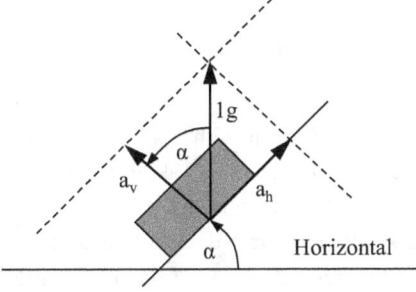

$$\sin \alpha = \frac{a_h}{1g} \qquad [38]$$

$$\alpha = \arcsin \frac{a_h}{1g} \qquad [39]$$

with α - Inclination angle
 a_h - Measured acceleration of the object in horizontal direction
 g - Gravity (9.81 m/s²)

Fig. 3.26: Distribution of gravitational acceleration on a stationary object on an inclined plane

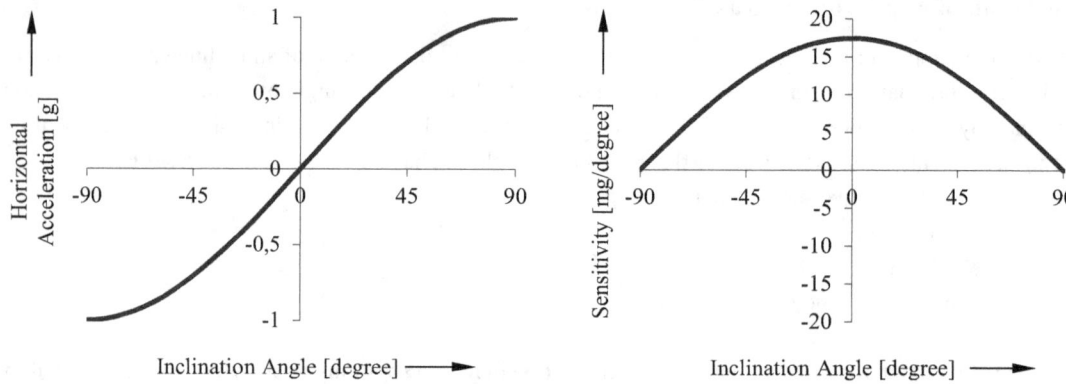

Fig. 3.27: Representation of the horizontal acceleration a_h as a function of the inclination angle and their sensitivity

For the zero point, the sensitivity is 7.5 mg/degree or invers expressed 0.057 degrees/mg. It follows that an acceleration measurement error of 1 mg results an inclination measurement error of 0.057°. The curve at 90° (i.e., in the vertical position) is different. Here is the sensitivity 0 mg/degree. This means that the measurement error in this area is infinitely large. For this reason, the inclination measurement with a single-axis acceleration sensor is only possible to a maximum of 60°. In addition to the flat curve and the resulting low sensitivity at large angles of inclination, the non-constancy of gravity plays a role as a source of error, which is due to the following causes[7]:

1. Latitude,
$$g_0 = 9.78 \, m/s^2 \cdot (1 + 5 \cdot 10^{-3} \sin^2 \Phi + 6 \cdot 10^{-6} \sin^2 2\Phi)$$

2. Height above the sea level,
$$g = g_0 - h \cdot 3 \cdot 10^{-6} \, s^{-2}$$

with g - *Acceleration due to gravity*
g_0 - *Acceleration due to gravity at sea level*
h - *Height above the sea level*
Φ - *Latitude*

3. Gravity anomalies due to mass inhomogeneities in the earth crust.
$$g = g \pm 4 \cdot 10^{-3} \, m/s^2$$

Geographical latitude has the biggest impact. Accelerometers perform on earth with a variation of the gravitational acceleration of 9.83 m/s² at the North Pole up to 9.78 m/s² at the equator. Taking the inverse sensitivity at 60° of 0.11 degrees/mg, a maximum measuring error of ± 0.3° is obtained by the variation of gravity field. In the zero point measurement (spirit level principle) this source of error does not occur. Here, the horizontal acceleration is equal to zero and is thus independent of the absolute value of the acceleration due to gravity.

---

[7] Kleine Enzyklopädie Natur, VEB Bibliographisches Institut Leipzig, 1983, p. 94

360°-Inclination Measurement with 2-Axis Acceleration Sensor

The 1-axis solution, in addition to the described non-existent sensitivity in the 90°-point, is also unsuitable for 360°-measurements, because the sensor has the same behavior in the range 0° to 90 ° and 90° to 180° as well as in the range of 180° to 270° and 270° to 360°. As a result, the measured values of the individual quadrants cannot be clearly assigned. This problem is solved by the use of a 2-axis acceleration sensor (a_x, a_y), which is rotated about its z-axis. The angle α results from the trigonometric equation [40]. The assignment of the measured values to the individual quadrants is possible via a signed integer comparison of a_x and a_y.

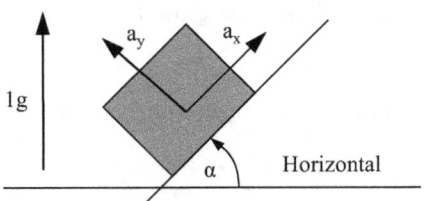

$$\alpha = \arctan \frac{a_x}{a_y} \qquad [40]$$

with α - Inclination angle
 a_x - Acceleration in x-direction
 a_y - Acceleration in y-direction
 g - Gravity

Fig. 3.28: 360°-measurement of the inclination angle by means of a 2-axis acceleration sensor

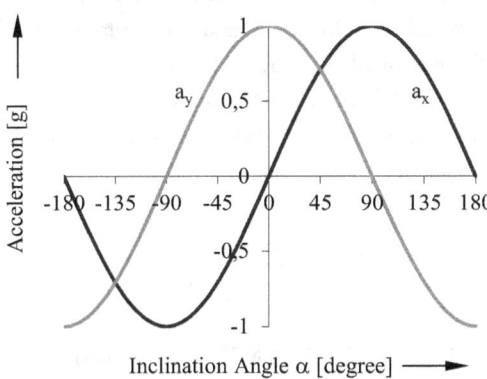
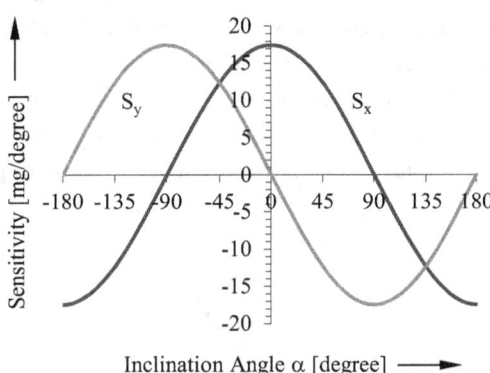

Fig. 3.29: Representation of the acceleration in x- and y-direction as a function of the inclination angle and their sensitivity

By using two sensors rotated by 90° to each other (x and y sensor), it is possible to keep the angular error relatively low over the entire 360° range. Fig. 3.30 shows this for a measurement error of the acceleration sensors of 1 mg.

$$|\Delta F_\alpha| = \left|\frac{\partial \alpha}{\partial a_x} \Delta a_x\right| + \left|\frac{\partial \alpha}{\partial a_y} \Delta a_y\right| \qquad [41]$$

$$|\Delta F_\alpha| = \frac{|a_y \cdot \Delta a_x| + |a_x \cdot \Delta a_y|}{g^2} \qquad [42]$$

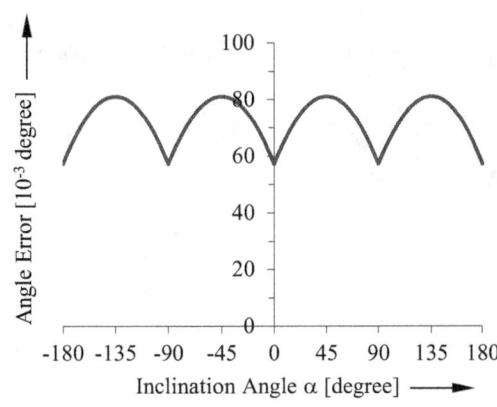

Fig. 3.30: Representation of the resulting error in angle determination due to erroneous acceleration measurement values
(Acceleration measurement error $\Delta a_x = \Delta a_y = 1$ mg)

For non-correlated influences, such as noise, the individual factors of the x- and y-axis must be summed up quadratically and one obtains:

$$\Delta F_\alpha = \frac{\sqrt{(a_y \Delta a_x)^2 + (a_x \Delta a_y)^2}}{g^2} \quad [43]$$

with
ΔF_α - Error of angle determination
Δa_x - Measurement error of the x-channel accelerometer
Δa_y - Measurement error of the y-channel accelerometer
a_x - Measured value of the x-channel accelerometer
a_y - Measured value of the y-channel accelerometer
g - Acceleration due to gravity

Thus a constant angular error (angular noise) of 0.057 deg over the entire angular range, with a measurement error of the accelerometers of $\Delta a_x = \Delta a_y = 1$ mg.

Inclination Measurement on an Inclined Plane

The rotational axis for tilting of the sensor does not always correspond to a coordinate axis of the measuring system. This case is often observed with mobile platforms, which should be aligned horizontally. In this arrangement the coordinate system of the sensor and the reference plane are tilted, so that there is no clear assignment of the measuring axes of both systems to each other. With a 2-axis acceleration sensor and Eq. [45], it is possible to determine the tilt-angle (α). If a 3-axis accelerometer is used, Eq. [46] can be applied. This has the advantage that it is independent of the size of gravity.

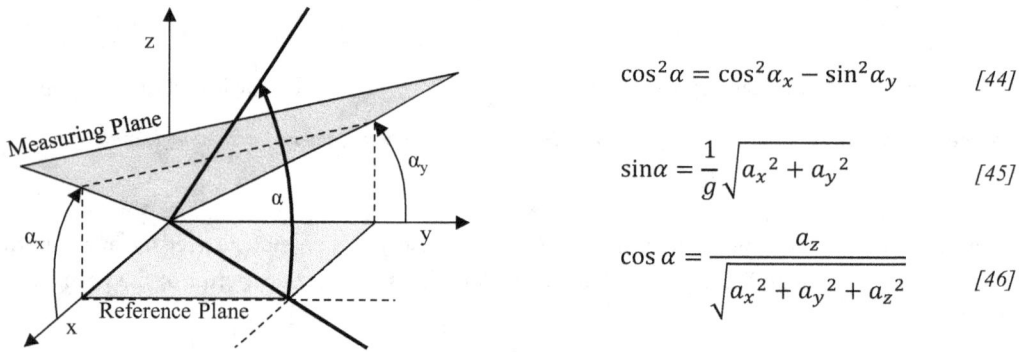

$$\cos^2 \alpha = \cos^2 \alpha_x - \sin^2 \alpha_y \quad [44]$$

$$\sin \alpha = \frac{1}{g}\sqrt{a_x^2 + a_y^2} \quad [45]$$

$$\cos \alpha = \frac{a_z}{\sqrt{a_x^2 + a_y^2 + a_z^2}} \quad [46]$$

Fig. 3.31: Measuring arrangement for determining the angle of inclination of a freely movable inclined plane

Application example: Condition Monitoring

The condition monitoring of machines is based on the continuous monitoring of machine parameters (e.g. power consumption, temperature), as well as measurable emissions (e.g., sound, vibrations, leaks). MEMS acceleration sensors are ideally suited for measuring vibrations (structure-borne sound).

For condition monitoring of machines, vibrations in the frequency range from 2 Hz to 1000 Hz are observed during normal operation. According to ISO 10816 [8], effective vibration speeds are defined in this frequency range, which must not exceed speeds of 2 mm/s to 11 mm/s, depending on the machine type, mounting method and age of the machine.

An inexpensive method for determining the vibration speeds involves measuring the accelerations on the machine and integrating the measured values over time.

$$v = \int a\, dt = \sum_{i=1}^{k} a_i \cdot \Delta t \qquad [47]$$

with v - Vibration speed
a_i - Measured acceleration
Δt - Time interval between two measurements

To avoid the integration of a DC component in the acceleration measurement (e.g., offset error, gravitational acceleration), this must be filtered out using a high-pass filter before integration. For the frequency bandwidth described above (2 Hz ... 1000 Hz) and a maximum vibration velocity measured to 11 mm/s, a necessary measuring range of the accelerometer of 69 m/s² is obtained.

$$\hat{a} = 2\pi f \cdot \hat{v} \qquad [48]$$

with $\hat{a}$ - Maximum acceleration
$\hat{v}$ - Maximum vibration speed
f - Frequency

This makes the commercially available 16 g accelerometers well-suited for condition monitoring of machines in accordance with the ISO 10816 standard. However, avoid violating the sampling theorem, it must be ensured that the upper limit frequency of the sensors is at least 2 kHz.

To monitor the wear of a machine's special components, it is also necessary to extend the frequency range up to 10 kHz. For example, a defective ball bearing with 10 balls emits a frequency ten times higher than the actual speed of the machine.

$$f_a = \frac{Z}{2} \cdot f_n \left(1 \pm \frac{d_K}{d_L}\right) \qquad [49]$$

with f_a - Emitted vibration frequency of the ball bearing
f_n - Relative velocity between the inner and outer ring of the bearing
Z - Number of balls
d_K - Diameter of the balls
d_L - Diameter of the bearing

---

[8] DIN ISO 10816-3: Mechanical vibrations - Evaluation of machine vibration by measurements on non-rotating parts - Part 3: Industrial machines with a nominal power over 15 kW and nominal speeds between 120 min⁻¹ and 15000 min⁻¹ when measured in situ

To distinguish the vibration components of the individual machine elements, it is advisable to break down the entire spectrum into the individual vibrations using FFT. In addition to the fundamental vibration (i.e., the actual rotation speed of the machine), a large amount of additional information is obtained that can be used to assess the state of wear of the machine. Fig. 3.32 shows such an image. The graph on the left shows the new condition of the machine. At a speed of 1200 rpm, a fundamental signal is obtained at 20 Hz. The second harmonic can also be seen. At -27 dB, the signal from the ball bearing hardly differs from the background noise in this measurement. It is different in the worn condition (right illustration). A clear increase in the signal around 200 Hz can be seen here, with harmonics and an increase in the general noise floor. Both are clear signs of wear and tear on the ball bearing and an imminent failure.

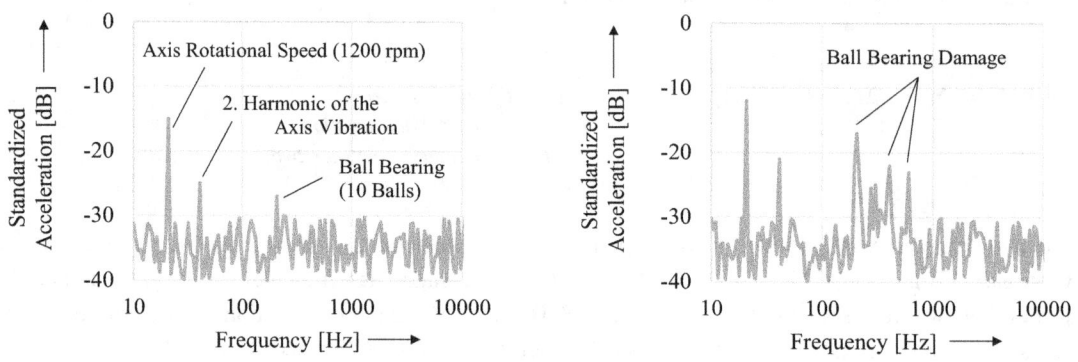

Fig. 3.32: Frequency spectrum of a rotating machine with and without a damaged ball bearing

MEMS acceleration sensors are typically offered up to a cut-off frequency of 10 kHz. For higher frequencies, microphones are used for condition monitoring. They can detect up to 100 kHz, which may be particularly necessary for lubrication condition monitoring. As high frequencies transmit well in air, microphones do not necessarily have to be attached directly to the machine. Hence, sensors with a combination of accelerometer and microphone are often found in these applications.

3.2 Gyroscopes

Gyroscopes, or strictly speaking angular rate sensors, measure the rotational speed of a body, the so-called angular rate [degree/s]. The angular rate is defined as the angle change per unit of time and is equal to the angular frequency (ω).

$$\omega = \frac{d\alpha}{dt} = 2\pi f = 360° \cdot f \qquad [50]$$

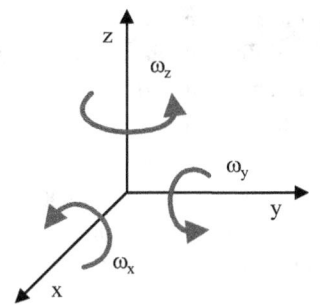

Fig. 3.33: Definition of the angular rates in space

Integrating the angular rate over time, the alignment of the body in space is received, described by the three Euler angles (φ, χ, ψ).

$$\phi = \int \omega_x dt \qquad [51]$$
$$\chi = \int \omega_y dt \qquad [52]$$
$$\psi = \int \omega_z dt \qquad [53]$$

with φ - Angle of rotation about the x-axis (roll angle)
χ - Angle of rotation about the y-axis (pitch angle)
ψ - Angle of rotation around the z-axis (yaw angle)

Angular rates can be measured very differently depending on the application. All methods are based either in mechanical systems on the measurement of a Coriolis force, or in optical systems on the Sagnac effect.

Optical Gyroscopes

The optical gyroscopes include laser and fiber gyroscopes. In both arrangements, two laser beams with opposite running directions are produced into a ring configuration. If the arrangement is rotated, the beam running in the direction of rotation must travel a shorter distance than the beam with the opposite orientation. The transit-time difference can be evaluated, and is proportional to the angular rate. It is advantageous that linear movements have no influence on the running time differences, and so laser gyros are insensitivity to it.

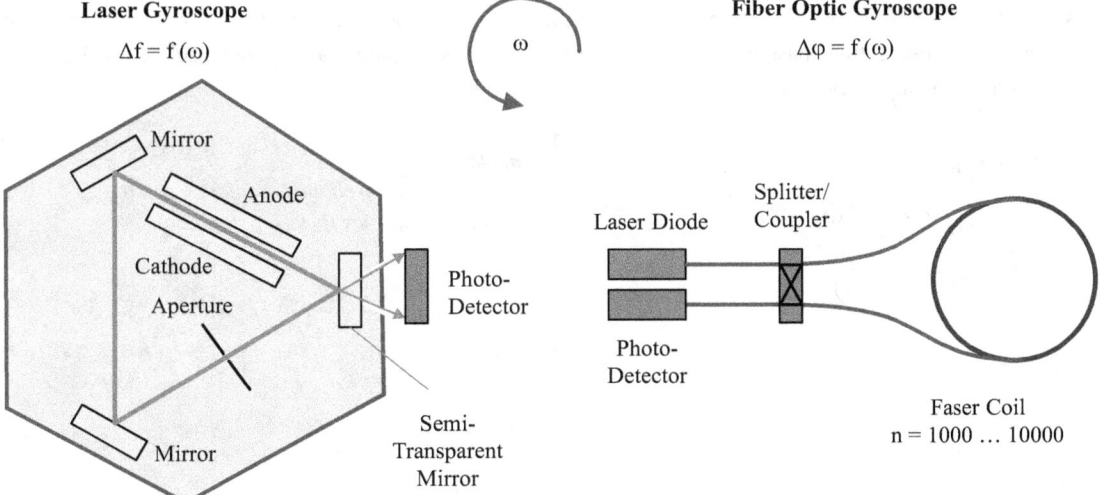

Fig. 3.34: Schematic representation of the structure of a laser gyroscope (left) and of a fiber optic gyroscope (right)

In laser gyroscopes, a laser beam is emitted in an evacuated channel, whose frequency is determined by the length of one circulation. If the laser gyroscope is applied with an angular rate, the circular length varies depending on the direction of rotation of the laser beam. The results are two frequencies (colors) of the laser beam, which differ from the original frequency at rest. Laser gyroscopes are high-precision gyroscopes with a typical zero stability of 0.01°/h.

Fiber optic gyroscopes are somewhat simpler. For them, a glass fiber winding leads the laser beam in a circle. A laser beam produced by a laser diode is passed through a splitter and sent in positive and negative direction of circulation through the glass fiber winding. If rotation occurs, there is a phase difference between the two laser beams, due to running time differences. This phase difference is detected on a photodetector by superposition of the two beams. Fiber-optic gyroscopes have typical a zero point stability of 1°/h.

MEMS Gyroscopes

MEMS gyroscopes are based on the principle of measuring the Coriolis force. Eq. [54] shows that the Coriolis force is proportional to the angular rate (ω), the mass (m) and the velocity (v) of the moving body. If the axis of rotation and the velocity of the moving body stand perpendicular to each other, the sin θ is equal one, and the relationship in Eq. [55] occurs.

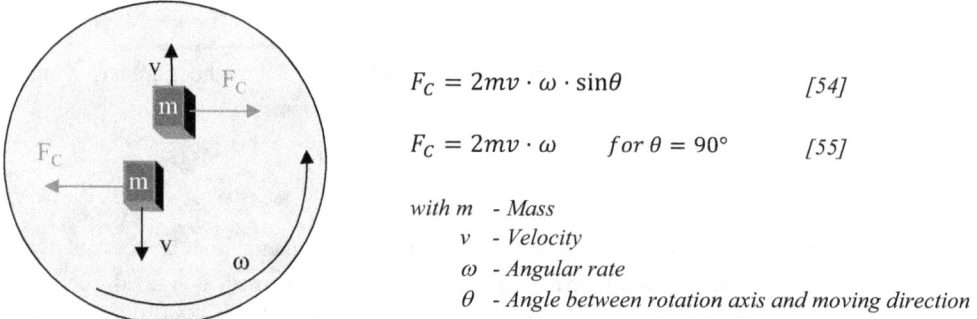

$$F_C = 2mv \cdot \omega \cdot \sin\theta \qquad [54]$$

$$F_C = 2mv \cdot \omega \qquad for\ \theta = 90° \qquad [55]$$

with m - Mass
v - Velocity
ω - Angular rate
θ - Angle between rotation axis and moving direction

Fig. 3.35: Coriolis force in a rotating system acting on a mass, which is moving on a straight line

Because translational movements are difficult to achieve in a millimeter-sized sensor chip, a vibrating mass is used as a source of motion (primary vibration). Through the constant change in the direction of movement of the oscillating mass a non-constant output value is obtained, a response oscillation. Eq. [56] describes the occurring Coriolis force, which causes the secondary vibration.

$$F_C = 2m\omega \cdot v_0 \cdot \sin(\omega_E t) \qquad [56]$$
$$for\ \omega \ll \omega_E$$

with ω - Angular rate
v_0 - Amplitude of the excitation oscillation speed
ω_E - Excitation frequency (primary vibration)

It is advantageous that the measurement results of gyroscopes, which are based on the Coriolis force principle, are independent of the distance of the center to the axis of rotation. For this reason, the sensor can be mounted at any place in the rotating system.

MEMS gyroscopes often consist of electrostatic excitations and capacitive readout elements. Both functional elements can be implemented in silicon using the same technologies. It is accepted, that a relatively large crosstalk of the excitation voltage on the capacitive measuring system occurs in this configuration. For very precise gyroscopes, sensors with piezoelectric excitation systems are also used, in which the crosstalk is significantly lower.

Fig. 3.36 shows the schematic structure of a MEMS gyroscope. In the center of the sensor is located the mass, which is excited by means of the electrostatic drive structures for the primary vibration in the y-direction. The oscillating mass is held in a frame and stimulated to a secondary vibration in x-direction when a rotation around the z-axis appears. This vibration is detected by a capacitive readout structure, as already discussed for acceleration sensors in the previous chapter.

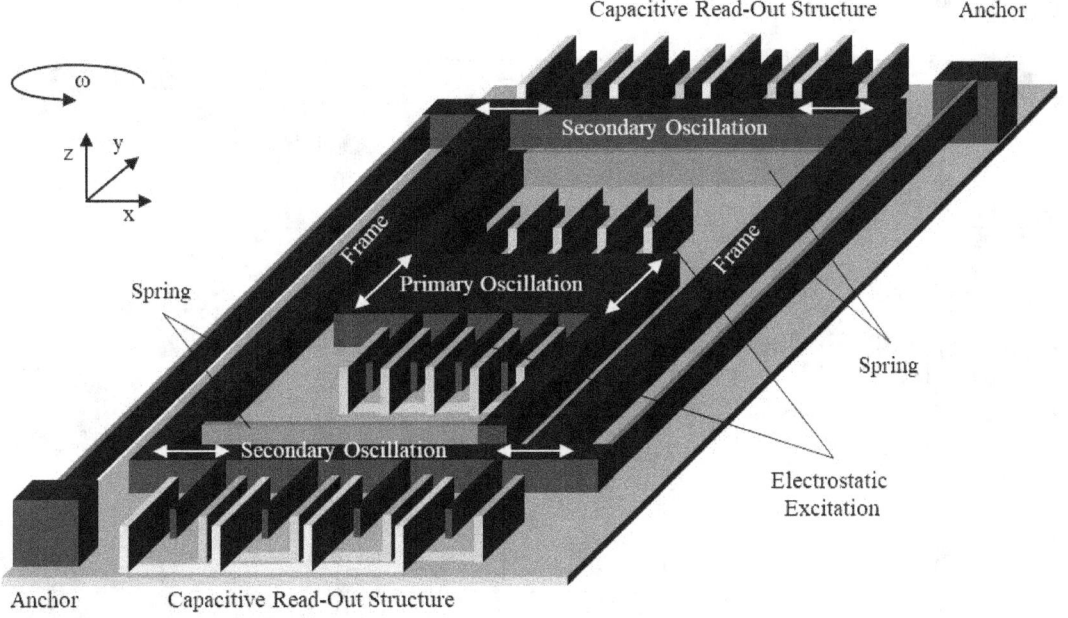

Fig. 3.36: Schematic representation of a MEMS gyroscope with electrostatic excitation and capacitive readout structure

Fig. 3.37 shows a chip photo of the 3-axis MEMS gyroscope L3GD20HTR[9]. It has both electrostatic excitation and capacitive readout structures.

The sensor's main drives can be seen on the left and right-hand sides of the picture. They excite a counter-rotating oscillation of the two masses in x-direction. The two masses in the y-direction (top and bottom) are excited by the main drive via the coupling springs at the corners of the structure, as well as by the two small drive structures at the top and bottom. The lower drive structure was destroyed during preparation and is therefore not visible in the image.

If a rotation around the z-axis occurs, the Coriolis force deflects the masses, so that an oscillation along the image plane occurs. This oscillation can be measured by the capacitive evaluation structures. For the detection of the other two spatial directions, the sensor structure contains electrode areas below the mass structures (not shown in the image). If there is a rotation around the x- or y-axis, a tilting movement out of the image plane occurs and can be detected with the electrode structures below the mass.

The sensor structure is 1.5 x 1 mm² and is hermetically sealed by a silicon lid. In addition to the sensor structure, the overall sensor contains an evaluation chip and an LGA interposer. The chip stack is finally plastic encapsulated and delivered as a 14-pin LGA package with the dimensions 3 x 3 x 1 mm³.

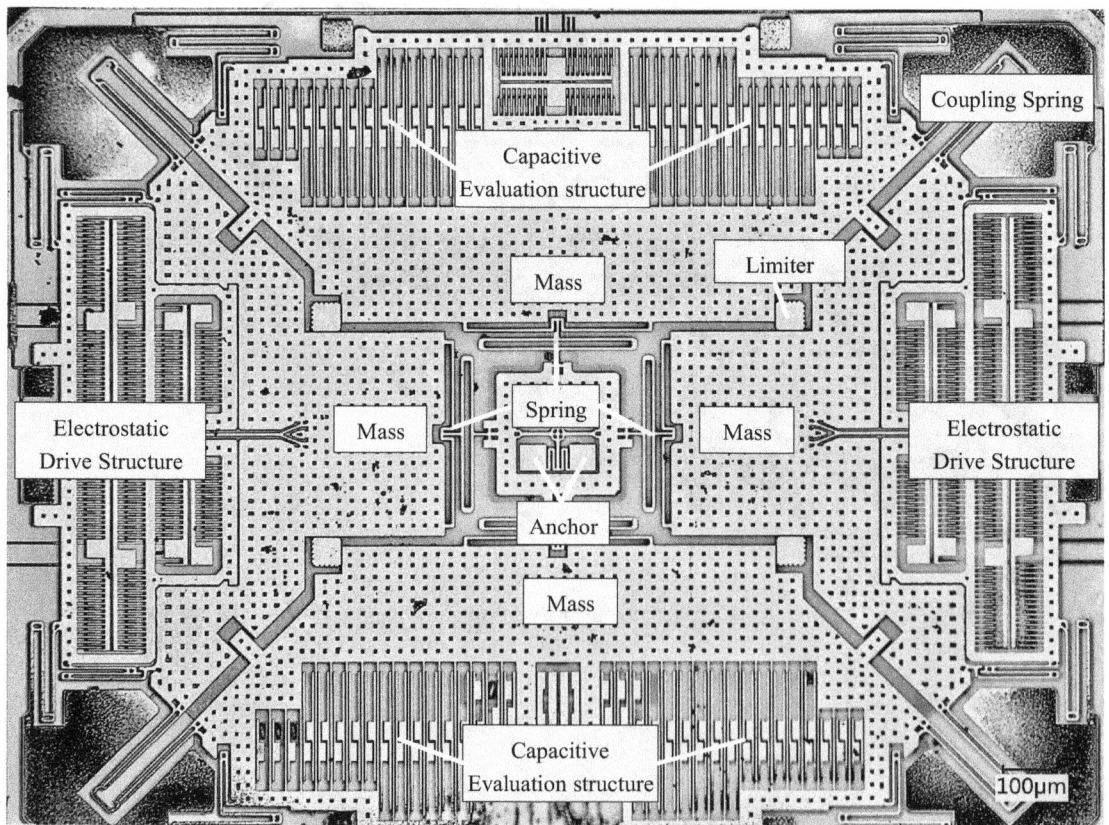

Fig. 3.37: Chip photo of the 3-axis MEMS gyroscope L3GD20HTR from STMicroelectronics.

---

[9] Datasheet L3GD20HTR, STMicroelectronics, 2013

Parameters of MEMS gyroscopes

In Tab. 3.5, the parameters of three MEMS gyroscopes are exemplarily listed. It is noteworthy that there is a higher power consumption compared to acceleration sensors. This is because gyroscopes require a mass to be actively excited to oscillate. Important parameters for gyroscopes are the stability of the zero point, the noise density and the sensitivity of the sensor to linear accelerations.

	ADXRS642 Analog Devices	BMG160 Bosch	BMI270 Bosch
Measuring Range	1-axis Gyro ± 300°/s	3-axis Gyro 125 - 2000 °/s	3-axis Gyro 125 - 2000 °/s 3-axis Acceleration 2 - 16 g
Sensitivity /Resolution	7 mV /°/s	16 bit	16 bit
Sensitivity Error		± 1 %	± 2 %
Non-Linearity	± 0.01 % FS	± 0.05 % FS	± 0.01 % FS
Noise Density	0.02 °/s/$\sqrt{Hz}$	0.014 °/s/$\sqrt{Hz}$	0.007 °/s/$\sqrt{Hz}$
Resonant Frequency	17 kHz		
Cut-off Frequency	until 2000 Hz	12 Hz … 230 Hz	25 Hz … 751 Hz
Cross-Axis Sensitivity		± 1 %	± 0.2 %
Zero-Point Bias	± 0.1°/s	± 1 °/s	± 0.5 °/s
Bias Temperature Coefficient	± 1 °/s/°C	± 0.015 °/s/°C	± 0.015 °/s/°C
Bias Stability	20 °/h		
Interface	Analog	SPI, I²C	SPI, I²C
Sensitivity to Linear Accelerations	0.03 °/s/g	0.1 °/s/g	0.1 °/s/g
Power Supply	5 V	2.4 … 3.6 V	1.7 … 3.6 V
Power Consumption (active)	3.5 mA	5 mA	685 µA
Package	BGA 7 x 7 x 3 mm³	LGA 3 x 3x 0.95 mm³	LGA 2.5 x 3 x 0.8 mm³
Release Year	2011	2014	2020

Tab. 3.5: Data comparison of different MEMS gyroscopes (typical values)

Offset Stability

The offset stability of a gyroscope describes the deviation of the angular value per unit of time when integrating the angular rate measured values over a period of time (τ). The measured values of the gyroscope are recorded at rest under constant ambient conditions (e.g., temperature). To determine the offset stability, the "Allan variance" is calculated for different integration times, which is defined as half of the average of the difference squares of two successive measured value mean values.

$$AV = \sigma^2(\tau) = \frac{1}{2}\frac{1}{(N-2m)}\sum_{k=1}^{N-2m}\left(\overline{\omega_{k+m}(\tau)} - \overline{\omega_k(\tau)}\right)^2 \qquad [57]$$

with $\tau = m \cdot t_a$

and $\overline{\omega_k(\tau)} = \frac{1}{m}\sum_{i=K}^{K+m-1}\omega_i$

with AV - Allan Variance
$\overline{\omega}$ - Mean value of the measured angular rate
ω_i - i^{th} angular rate measured value
τ - Period of averaging
N - Total number of measurement values
m - Number of measured values for averaging
t_a - Sampling time

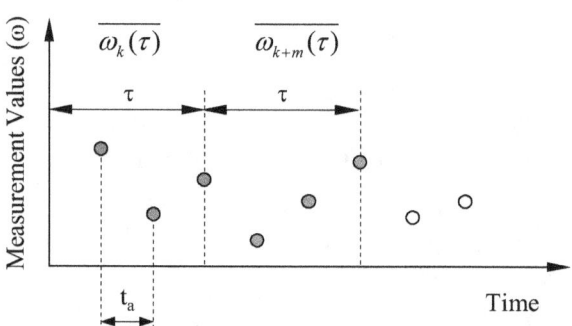

Fig. 3.38: Determination of the mean values for the calculation of the Allan variance

If one takes the square root of the Allan variance, one obtains the Allan deviation [grad/s].

$AD = \sqrt{AV}$ [58]   with AV - Allan-Variance
                                           AD - Allan-Deviation

If the Allan deviation is determined for different mean value intervals (τ) and plotted double-logarithmically in a diagram, the Allan deviation plot is obtained. The minimum of the plot represents the sensor's offset stability, which describes the dispersion of the angle determined from the noisy rotation rate measurements by integration for an ideal averaging interval.

Fig. 3.39 shows the Allan plot for the gyroscope ITG320014. The sampling frequency of this measurement is 50 Hz. The first part of this graph is defined by the noise level. If the noise is completely non-correlated, a straight line with a slope of - 0.5 is seen in this part of the graph. The minimum of the graph represents the bias stability of the sensor. In this case, it is 0.0012 deg/s. Converted into the common unit [deg/h], this result in a value for the offset stability of 4.3 deg/h. At the end, the graph goes up, caused by inherent instabilities in the output of the sensor.

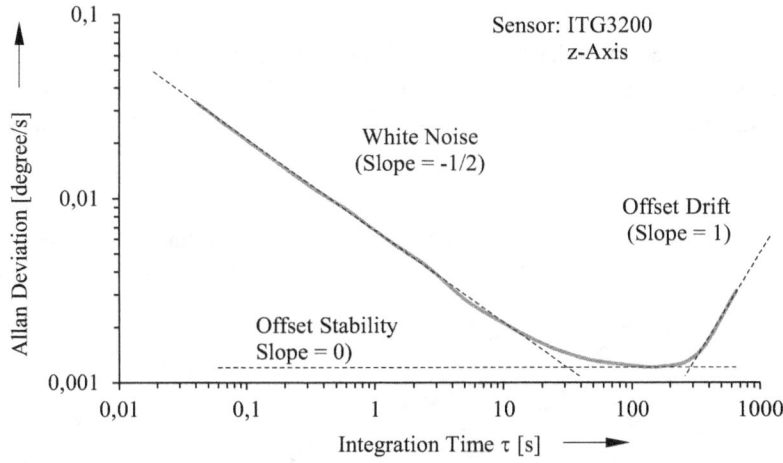

Fig. 3.39: Allan Deviation Plot for the gyroscope ITG3200 [10]

For sensor data that are integrated, the noise not only has a momentary uncertainty for the measurement values, but also a scattering of the integral. This scattering is called "Radom Walk". If the value from the Allan deviation plot is multiplied by the respective integration time, the deviation of the angle values at this integration time is obtained.

Fig. 3.40 represents such a course for ten random integrated intervals, of measurements data of a gyroscope.

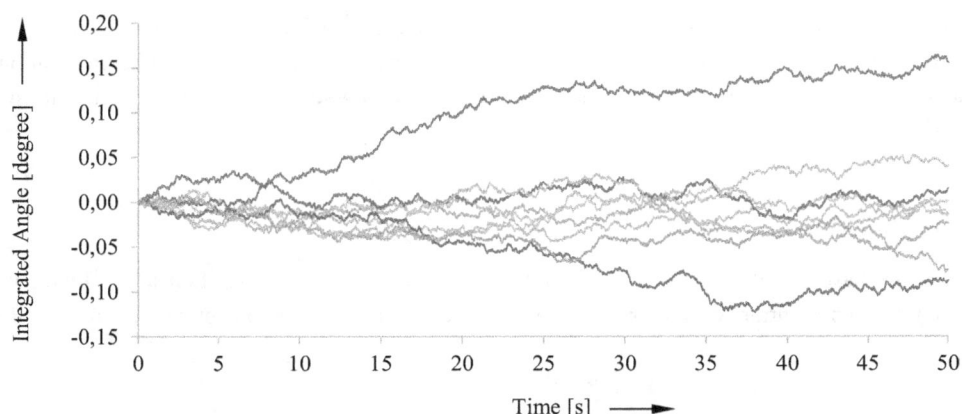

Fig. 3.40: "Random walk" of the gyroscope IT3200 [10] at rest with an effective noise of 0.05 degree/s

---

[10] Datasheet ITG3200, InvenSense Inc., 2010

Applications of Gyroscopes

Gyroscopes are widely used in applications where a rotational movement should be measured, but the rotation axis is not known or metrological inaccessible. Fig. 3.41 shows the use of gyroscopes in the automobile with the respective required resolutions. Other areas of application are for example the control of the tilt angle in tilting trains or balancing unicycles (e.g., Segway).

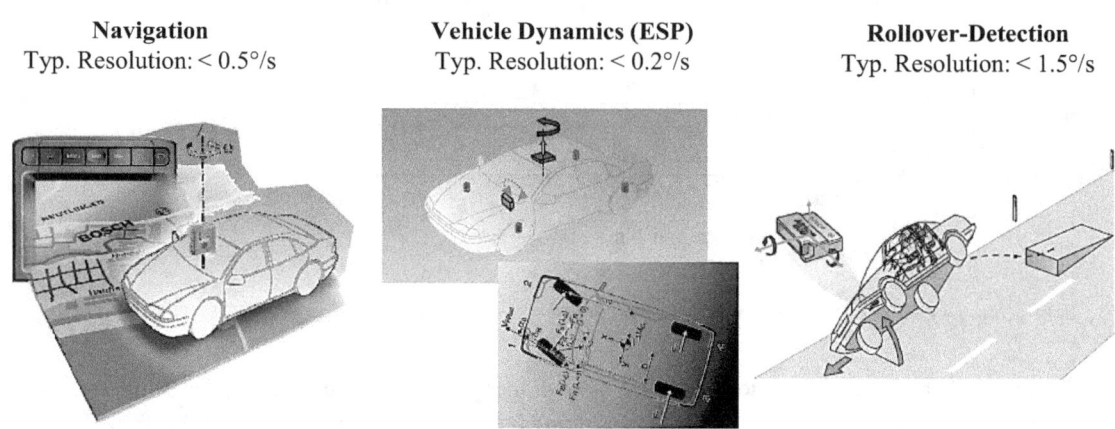

| **Navigation** | **Vehicle Dynamics (ESP)** | **Rollover-Detection** |
Typ. Resolution: < 0.5°/s — Typ. Resolution: < 0.2°/s — Typ. Resolution: < 1.5°/s

Fig. 3.41: Use of gyroscopes in cars (Source: Bosch)

However, initially the automotive industry was the main customer for MEMS gyroscopes, smartphone manufacturers are now the main customers. Concomitantly, there could again be a significant reduction in the dimensions and power consumption of the gyroscopes. In most cases, the gyroscopes in smart phones are used only for interactive motion detection. However, they are also powerful enough to perform more complex tasks, such as inertial navigation. For such applications, magnetic field sensors can also be integrated, in addition to gyroscopes and acceleration sensors, to get a better signal quality through data fusion of the signals from the individual sensors.

Inertial Navigation Systems (INS)

INS-systems are based on the measurement of the angular rate and the acceleration in all three spatial directions and the determination of the orientation and the position of an object by integrating these measurements.

$$x = \iint a_x \, d^2 t \quad [59] \qquad \varphi = \int \omega_x \, dt \quad [60]$$

$$y = \iint a_y \, d^2 t \quad [61] \qquad \chi = \int \omega_y \, dt \quad [62]$$

$$z = \iint a_z \, d^2 t \quad [63] \qquad \psi = \int \omega_z \, dt \quad [64]$$

If an accelerometer is hung on gimbals and the rotational movements is regulated so that the resulting rate of rotation is in all three spatial directions is zero, one obtains a very simple mathematical INS-system. Because in this case the sensor coordinate system always matches to the outer reference system, and so it is possible to determine the position about a double integration of the acceleration. Only the gravitational acceleration must be subtracted from the z-value of the measured acceleration.

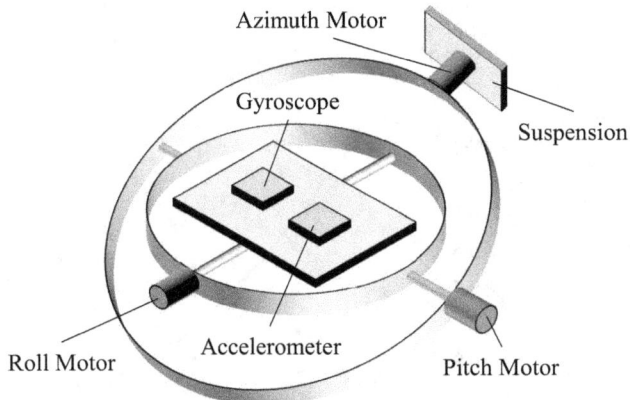

Fig. 3.42: INS system with stabilized platform

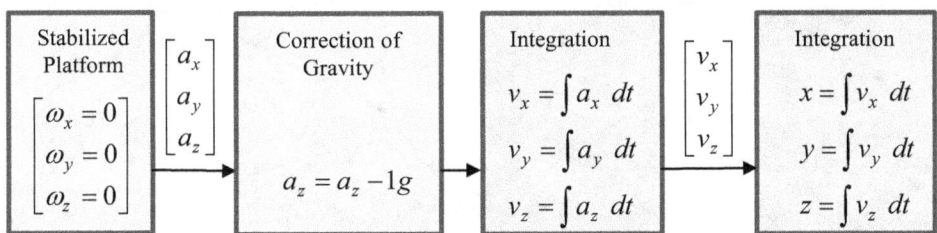

Fig. 3.43: Solution algorithm of an INS-System with a stabilized platform

Fig. 3.44 shows an example of position determination by means of a stabilized sensor platform. There are given the acceleration measurement values (right diagram) for a movement at a constant speed (v_0) on a circular path. By being stabilized, the alignment of the sensor does not change in spite of the circular motion. For reasons of clarity, a 2-dimensional representation is used, ignoring the z-component. The gravity correction can therefore be omitted. In the two lower diagrams in Fig. 3.44, the results of both integrations are shown. It is evident that for error-free measured values the location determination with the integration of the measured acceleration values even for non-linear movements works perfectly.

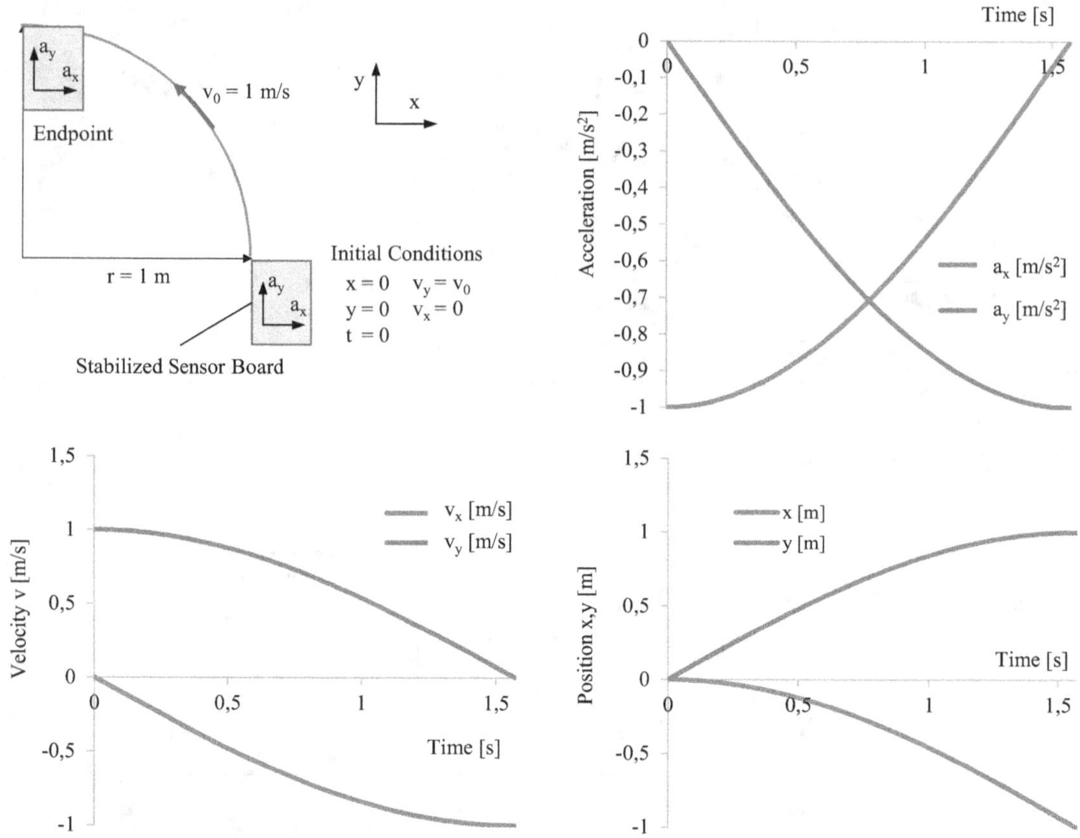

MEMS

Fig. 3.44: Example of the measurement by means of stabilized sensor platform for the motion on a circular path

Gimbals with a gyroscopic adjustment are very cost- and space-intensive. Nevertheless, they are used for simple systems. Fig. 3.45 shows a gimbal suspension of an electronic compass, which always remains in a horizontal position due to an additional weight at the bottom of the suspension.

Fig. 3.45: Passive gimbal suspension of an electronic compass of a self-steering system for sailboats

In order to be able to do without the complex gimbal suspension, a mathematical solution is often used. Fig. 3.46 shows the flow chart of such an approach. It is much more complex than the approach with a stabilized platform, since the rotation of the sensor changes the orientation of the measurement and the reference coordinate system to each other. The required coordinate transformation for settling of measurement and influence factors is marked with the box >rotation<. The formulas behind this box are given in Eq. [65]-[70] and Tab. 3.6. Due to the rotational transformation of the acceleration, the gravity no longer acts exclusively on the z-axis in a twisted sensor. There are also effects of gravity, which produce a signal on the x- and y-axis. The reverse transformation is necessary in the speed measured from the sensor. Here, the true direction of movement in relation to the external reference coordinate system must be determined using the rotation matrix. Furthermore, it must be noted with this approach that in addition to the compensation of the acceleration due to gravity, the centrifugal acceleration must also be compensated mathematically.

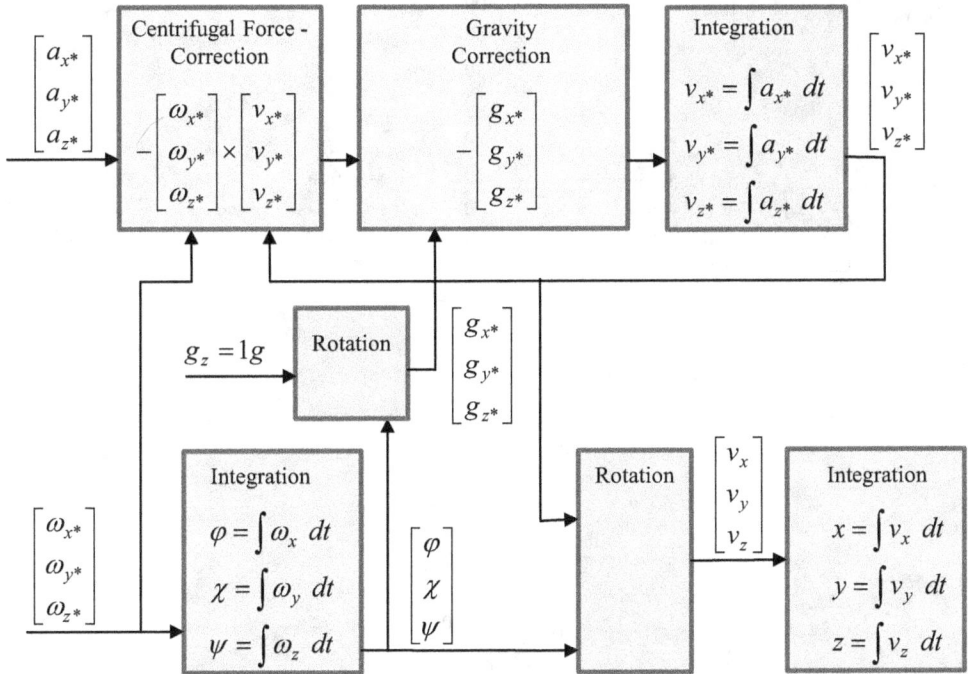

Fig. 3.46: INS-system with non-stabilized platform

$v_x = a_{11}v_x^* + a_{12}v_y^* + a_{13}v_z^*$ [65] $g_{x*} = a_{31}g_z$ [66]

$v_y = a_{21}v_x^* + a_{22}v_y^* + a_{23}v_z^*$ [67] $g_{y*} = a_{32}g_z$ [68]

$v_z = a_{31}v_x^* + a_{32}v_y^* + a_{33}v_z^*$ [69] $g_{z*} = a_{33}g_z$ [70]

a_{ik}	k = 1	k = 2	k = 3
i = 1	cos χ cos ψ	- cos χ sin ψ	sin χ
i = 2	cos φ sin ψ + sin φ sin χ cos ψ	cos φ cos ψ - sin φ sin χ sin ψ	- sin φ cos χ
i = 3	sin φ sin ψ - cos φ sin χ cos ψ	sin φ cos ψ + cos φ sin χ sin ψ	cos φ cos χ

Tab. 3.6: Coefficients of the rotation matrix

Fig. 3.47 shows the same example as in Fig. 3.44. The sensor platform is not stabilized, and rotates while moving on the circular path. A constant angular rate of 57.3 degrees/s is measured, resulting in a change of the angle of 90° about the integration time of 1.5 seconds. Furthermore, one measures a centrifugal acceleration of -1 m/s² in the x- direction, during the movement along the circular path. Since this is eliminated by means of the centrifugal acceleration correction, no change of the velocity components occurs. In this example it can be clearly seen that the sensor coordinate system no longer corresponds to the original coordinate system at the end of the circular path.

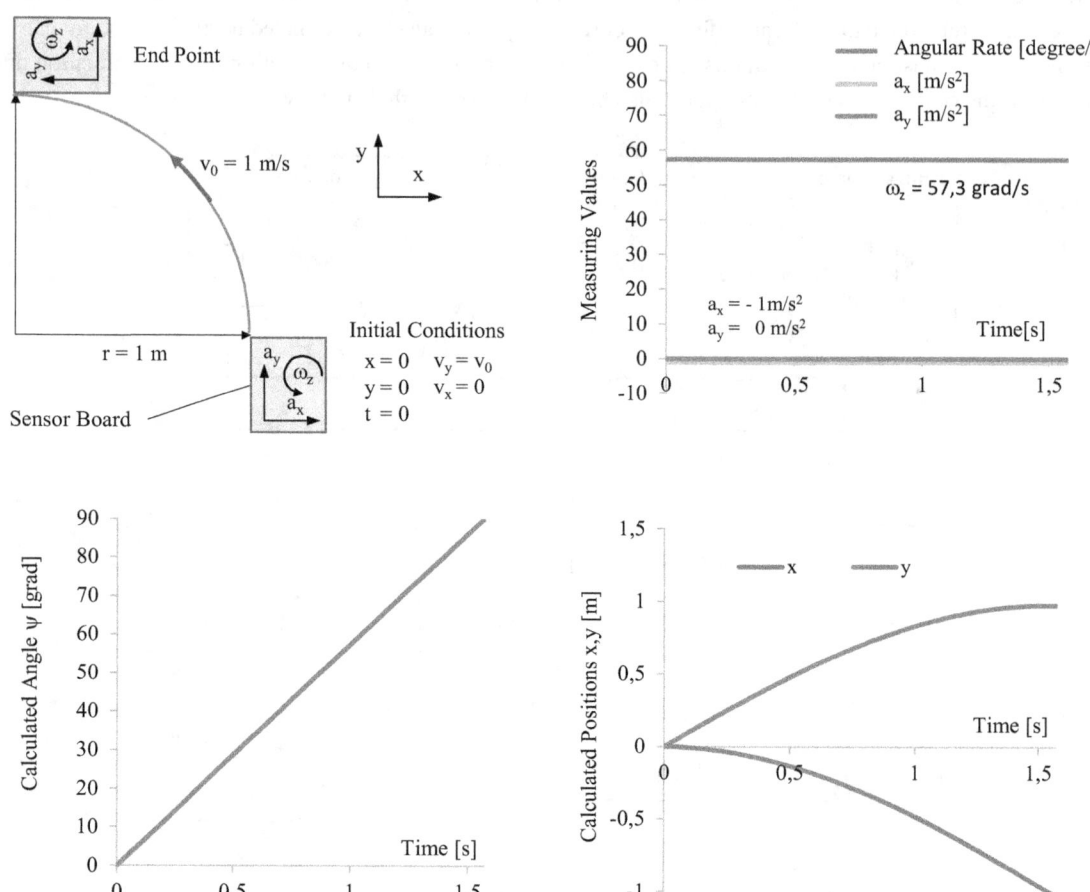

Fig. 3.47: *Example of the measurement by means of a non-stabilized sensor platform for the motion on a circular path*

Due to the manifold necessary compensations of this system, the determination of an exact location can be carried out only for a relatively short time interval. The main factor for the uncertainty of measurement is the gyroscope. If an error in the determination of the orientation occurs, the gravitation is incorrectly compensated and the relatively high acceleration value of 1g is divided incorrectly on the axes. This results, after the double integration of the acceleration values, in a large measurement error for the position determination. For this reason, it is important for such INS-approaches to keep the measurement errors of the sensors as small as possible. One way to accomplish this is the calibration of the sensors as well as the compensation of the temperature response. Another approach is the use of additional sensor data (magnetic field, visual or GPS) to build a self-calibrating system by using sensor data fusion.

3.3 Pressure Sensors

Pressure is generally understood as the force per unit area applied through gaseous or liquid medium to a solid. Traditionally, one finds for pressure a plurality of units.

$$p = \frac{F}{A} \qquad [71]$$

1 Pa	$= N/m^2 = kg \cdot m^{-1} \cdot s^{-2} = 10^{-5}$ bar	(Pascal)
1 at	$= 1 \text{ kp/cm}^2 = 98.066$ kPa	(Technical atmosphere)
1 atm	$= 101.325$ kPa	(Physical atmosphere)
1 psi	$= 1 \text{ lb/in}^2 = 6895$ Pa	(Pounds per square inch)

Pressure sensors consist of a membrane and at least one cavity. If a pressure difference is applied to the membrane, the membrane deflects and produces a measurement signal proportional to the pressure. The deflection of the diaphragm is detected either capacitively or piezo-resistively with MEMS pressure sensors.

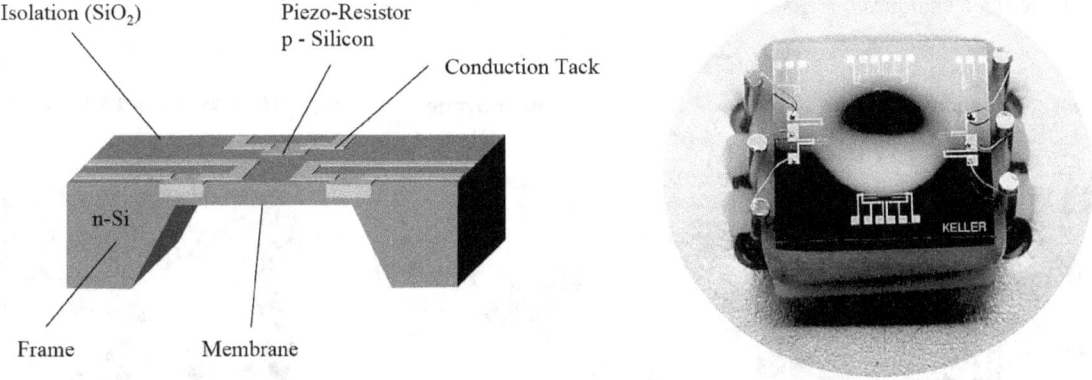

Fig. 3.48: Cross-sectional representation and picture of a piezo-resistive pressure sensor (Source: KELLER AG, www.keller-druck.com)

Fig. 3.48 shows the schematic structure of a MEMS pressure sensor with a piezo-resistive evaluation. The frame and the membrane are usually made of monocrystalline silicon, which was thinned by orientation dependent etching in the membrane area. In the region of the bending point of the membrane, where the greatest expansion of the material occurs, the piezo-resistors (shown in orange) are diffused. These are electrically contacted via a conductor track (shown in gray). On the right side of Fig. 3.48, a pressure sensor is shown. It is good to see the deflection of the diaphragm due to the applied pressure.

One way to classify pressure sensors is the subdivision after the membrane material. Here the silicon membrane of monocrystalline material and diffused piezo-resistive resistors are the standard solution. In addition to the excellent mechanical properties of silicon, the material is also characterized by the possibility of integrating the evaluation electronics onto the silicon chip. In the case of a cost-effective solution, the monocrystalline silicon is replaced with polycrystalline membrane material. This saves the expensive orientation-dependent etch process, and the sensor can be completely prepared in a standard CMOS manufacturing line. It is seen in Tab. 3.7 that due to the poor material properties and the limited thickness variation possibilities, the pressure range of polysilicon is greatly restricted. Polysilicon pressure sensors are mainly used for microphone applications.

MEMS

Membrane Material	Sensor Principle	Typ. Pressure Range [bar]
Silicon	Piezo-resistive (bulk-micromechanics)	0.01 ... 400
Poly-Si	Capacitive, piezo-res. (surface-micromechanics)	0.5 ... 10
Ceramic	Capacitive	0.001 ... 300
Ceramic	Thick-/ Thin-film technology, piezo-resistive	0.1 ... 300
Steel	Thin-film (metal or Poly-Si) piezo-resistive	1 ... 5000

Tab. 3.7: Overview of possible membrane materials with their respective pressure ranges

Silicon membranes have difficulties at high pressures, and especially at high pressure gradients, because silicon is a very brittle material and its toughness is very low. For applications of pressures p > 300 bar steel membranes are mainly used. Fig. 3.49 shows such a membrane, on which piezo-resistive elements are applied by sputtering techniques.

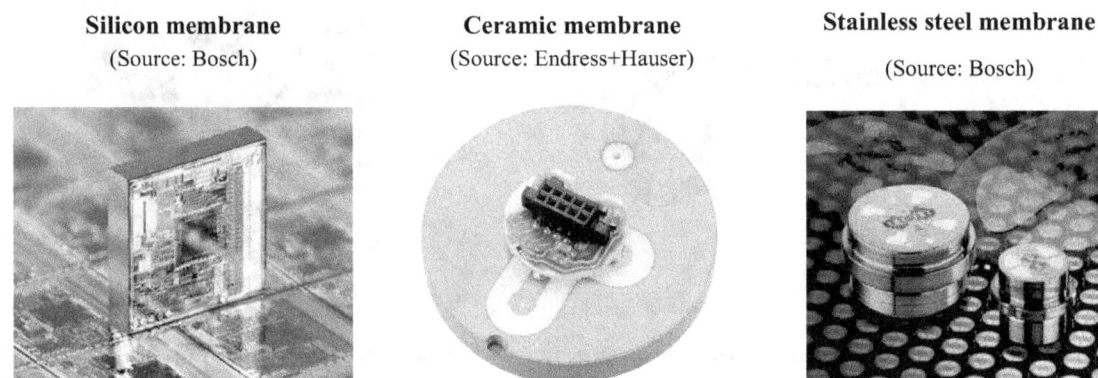

Fig. 3.49: Pressure Sensors with various membrane materials

The packaging technology play a decisive role in pressure sensors. First, it must be ensured that as little stress as possible is transmitted from the frame system to the sensor membrane. However, the sensor must be physically connected to the pressure media. Pressure sensor with metal package (TO8) (Source: Bosch) shows a typical metal housing from the early days of the pressure sensor. It is good to see in this example the glass (Pyrex) socket, on which the silicon chip is mounted. This socket minimizes the stress on the silicon that is caused by the assembly process and the differences in expansion of metal and silicon during temperature changes. The connection to the medium to be measured is accomplished by a metal tube, which is hermetically sealed by a glass lot. Because the metal lid is hermetically welded, there is the possibility to set a reference pressure in the metal housing. Despite all the advantages of such housings, today mostly plastic or ceramic housings are used due to financial reasons.

Fig. 3.50: Pressure sensor with metal package (TO8)
(Source: Bosch)

Hybrid packages are an intermediate step towards fully integrated plastic packages. They are based on the use of a base substrate, onto which the pressure sensor and the electrical and pneumatic connections are subsequently mounted. Either printed circuit board or ceramic substrates are used for the base substrate.

Fig. 3.51: Pressure sensor with hybrid housing, closed and open (KELLER AG)

Since a hybrid structure is not hermetically sealed, a reference pressure, as required for an absolute pressure measurement, must be provided in the sensor itself. For this purpose, either a cavity is created in the silicon by bonding a cover (Fig. 3.52) or a hermetically sealed cavity is already formed in the silicon during the manufacturing process (see Fig. 3.54).

Fig. 3.52: Pressure sensor chip with glass cover
(KELLER AG)

In industrial applications a surrounding housing, in addition to the sensor package, is often still used, which simplifies the installation, provides the connecting wires and protects the sensor against adverse environmental conditions (shock, humidity, dust). Therefore, a sensor arises from the primary microsystem to overall dimensions in the cm range. Fig. 3.53 illustrates the assembly of such sensor, which is designed for high pressures.

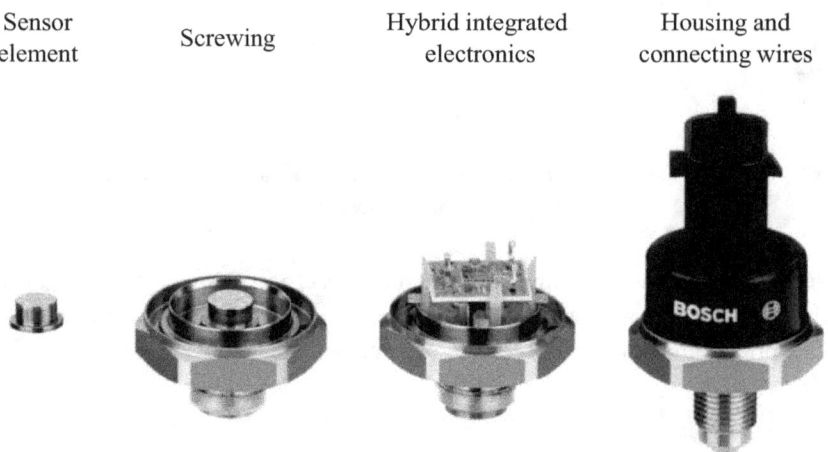

Fig. 3.53: Assembly of a pressure sensor for a fuel injection system in automobiles (Source: Bosch)

A further step towards the cost-effective production of pressure sensors and thus, a prerequisite for the use of pressure sensors in smartphones, is the further miniaturization of the housing by means of plastic molded LGA housings. To this end, three conditions must be particularly observed for pressure sensors. Firstly, it is necessary to provide pneumatic inlets in the LGA package. Furthermore, it must be taken into account that plastic encapsulation is not gas-tight, and therefore, not suitable for providing the reference pressure. Furthermore, a plastic encapsulation exerts a relatively high mechanical stress on the pressure sensor because it is in direct contact with the plastic encapsulation on all sides.

Fig. 3.54 shows an approach used in the LPS25HB[11] absolute pressure sensor. For the gas inlet, the lid of the silicon chip provides openings, which are positioned flat on the upper side of the LGA housing. The cavity for providing the reference pressure is located underneath. It is integrated directly into the silicon chip using sacrificial layer technology.

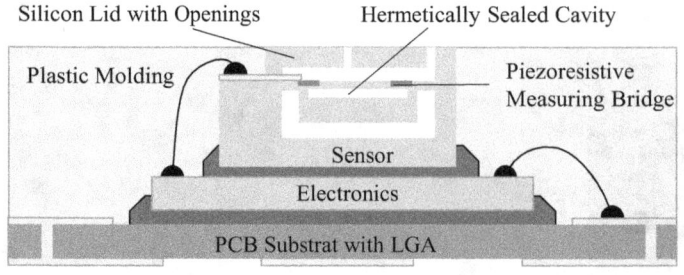

Fig. 3.54: Structure of a pressure sensor in a plastic-molded LGA housing

---

[11] Datasheet LPS25HB, STMicroelectronics, 2016

To minimize the stress on the piezoresistive sensor elements caused by the plastic encapsulation, this packaging design also integrates a silicon suspension in the sensor structure. Fig. 3.55 shows two of these variants. In the figure on the left, the design consists of four polygon springs that mechanically decouple the membrane surface from the frame. Each polygon spring contains two conductive tracks that are used to connect the piezoresistive resistors of the sensor bridge. In the right figure, the membrane surface is suspended by only one silicon arm. The eight leads, connected to the piezoresistive resistors along the silicon arm, are clearly visible in this design.

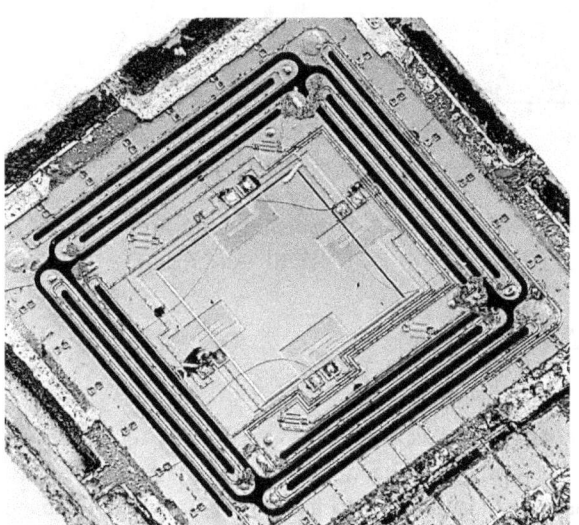 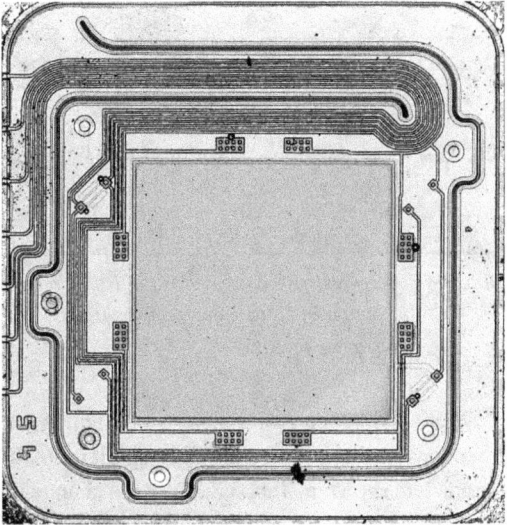

Fig. 3.55: Piezoresistive pressure sensors with integrated stress-minimizing suspension (STMicroelectronics LPS25 and LPS22)

Piezo-Resistive Evaluation

The piezo-resistive evaluation is, according to the capacitive evaluation, the most commonly used method for measuring deflections of micromechanical structures. It is based on the principle, that conductive materials change their resistance if they are stretched or compressed.

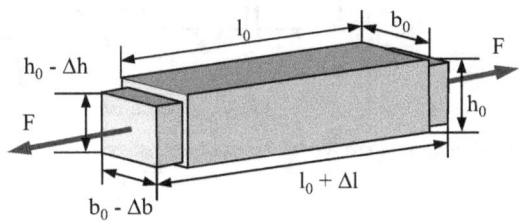

$$\frac{\Delta R}{R} = K \cdot \frac{\Delta l}{l_0} \qquad [72]$$

$$\frac{\Delta R}{R} = K \cdot \frac{F}{A} \cdot \frac{1}{E} \qquad [73]$$

with K - Piezo-resistive coefficient
 E - Young's modulus

The change in resistance in metallic materials is entirely due to the changes in the geometry of the component. One obtains a significantly larger piezo-resistive effect in semiconductor materials. This is based on the change of the band structure of the semiconductor, caused by the compression or stretching of the atomic lattice. To describe the size of the piezo-resistive effect, the piezo-resistive coefficient is used. It is the proportionality coefficient, which describes the percentage change in resistance at a percent change in length of the component. If the piezo-resistive coefficient in metallic materials lies in the range of two, it lies for silicon up to 150.

Material	K_L	K_T	α [10^{-3} K^{-1}]
n-Si	-52.7	-29.7	
p-Si	121.3	-112.1	0.5 ... -7.5
Poly-Si (boron doped)	30		

with K_L - Piezo-resistive coefficient for elongations in the current direction
K_T - Piezo-resistive coefficient for elongations perpendicular to the current direction
α - Temperature coefficient of the resistance

Tab. 3.8: Piezo-resistive coefficients for 100-silicon with current direction in [110][12] and temperature coefficient of resistance for silicon

Actually, the piezo-resistive coefficient is a tensor, which reflects the change in resistance in three spatial directions at force effects in three spatial directions. For better understanding, this tensor is reduced to two parameters shown in Tab. 3.8. Only the change in resistance for stretching the material in the current direction (K_L) and perpendicular to the current direction (K_T) is considered.

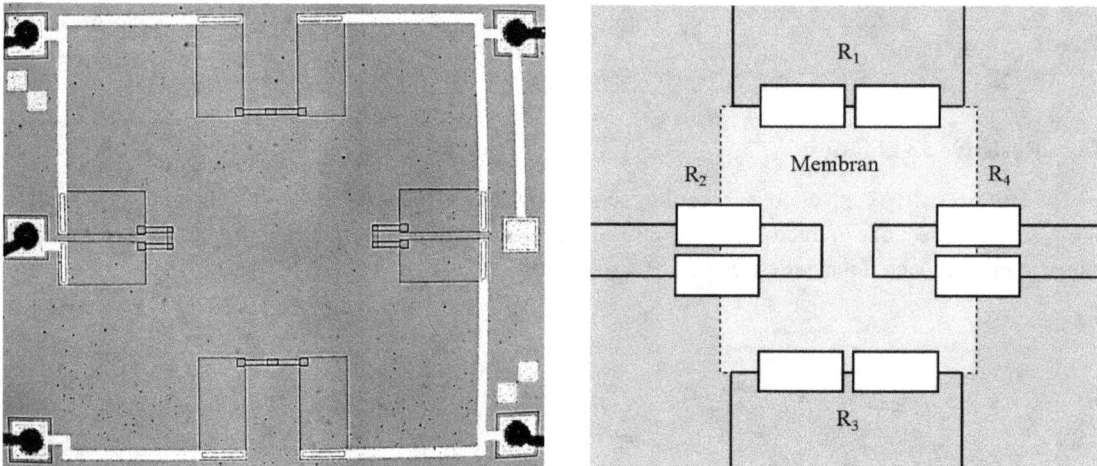

Fig. 3.56: Top view of a pressure sensor and schematic representation of the arrangement of piezo-resistive resistors on the sensor diaphragm

Fig. 3.56 shows the arrangement of the piezo-resistive resistors on the silicon chip. The position of R_1 and R_3 are selected such, that at an elongation of the membrane acts perpendicular to the current flow direction, whereas for R_2 and R_4, the elongation occurs in the direction of the current flow.

---

[12] J. Frühauf, Werkstoffe der Mikrotechnik, 2005, S.127, ISBN 3-446-22557-9

Readout of Measuring Bridges

For the evaluation of the resistance change in piezo-resistive pressure sensors, a measuring bridge (see Fig. 3.57) is recommended. By the special arrangement of the resistive sensor elements, a simple resistance-voltage relationship arises, following Eq. [74]. At the zero point, no temperature dependence occurs despite of the high temperature coefficient of silicon, because the temperature response of the individual resistances are compensated in the bridge arrangement.

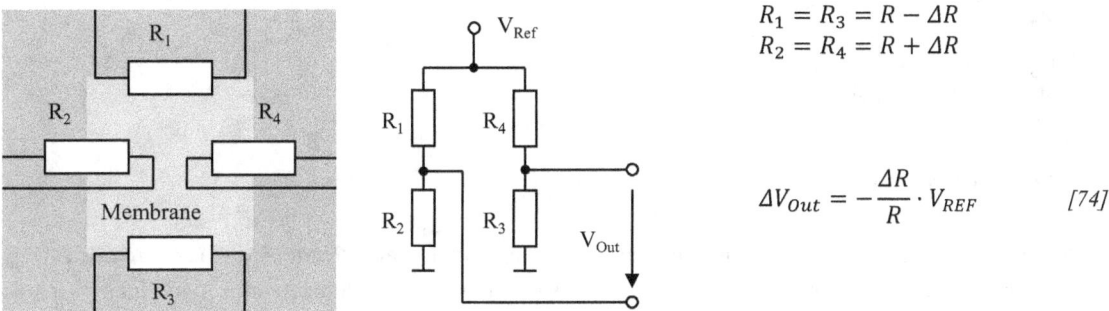

$$R_1 = R_3 = R - \Delta R$$
$$R_2 = R_4 = R + \Delta R$$

$$\Delta V_{Out} = -\frac{\Delta R}{R} \cdot V_{REF} \qquad [74]$$

Fig. 3.57: Top view of a piezo-resistive pressure sensor and wiring of the resistors to a measuring bridge

To amplify the differential output voltage (V_{Out}) of the measuring bridge, a differential amplifier can be used according Fig. 3.58. The amplifier picks up the signal with a high impedance, and converts it into a mass-based signal (single ended). If required, the signal can be additionally amplified by the dimensioning of the resistors R_5 and R_6. It should be noted, that the relatively low differential voltage at the bridge output are opposite to a high common mode voltage. In order to minimize the error influence of the high common mode voltage, an amplifier with a high common mode rejection ratio should be selected. Furthermore, the resistors R_5 and R_6 have to be much larger as the bridge resistors, to avoid the load of the bridge signal.

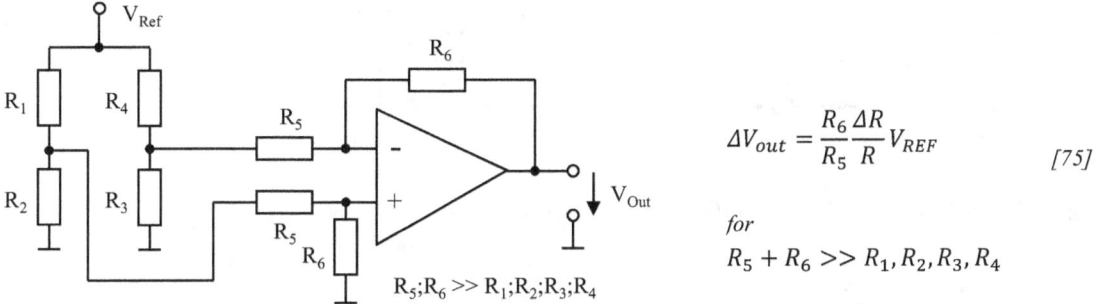

$$\Delta V_{out} = \frac{R_6}{R_5} \frac{\Delta R}{R} V_{REF} \qquad [75]$$

for
$$R_5 + R_6 \gg R_1, R_2, R_3, R_4$$

Fig. 3.58: Differential amplifier for reading out the measuring bridge

Another readout circuit is shown in Fig. 3.59. It has the advantage that an occurring offset voltage of the sensor may be corrected by R_{Tr}. Due to the possible single-ended voltage tap-off, a simple inverting amplifier can be used with a very high input impedance.

$$U_A = \left(1 + \frac{R_5}{R_6}\right)\left(\frac{\Delta R}{R}V_{REF} + V_{off}\right) \qquad [76]$$

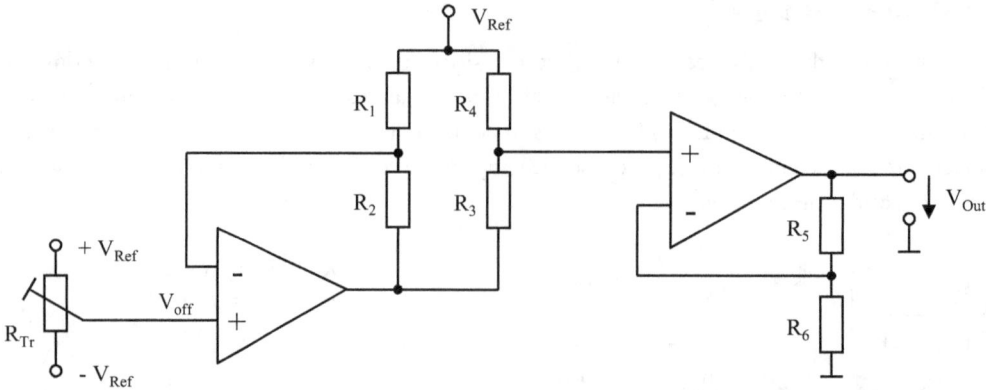

Fig. 3.59: High impedance readout circuit with offset correction

For the direct digitalization of bridge voltages, special analog-digital converter (ADC) are provided. They are characterized by a differential input with a very small input voltage range and a high common mode rejection ratio (CMRR). Optionally, these circuits have a constant current source for driving the measuring bridge as well as a temperature sensor.

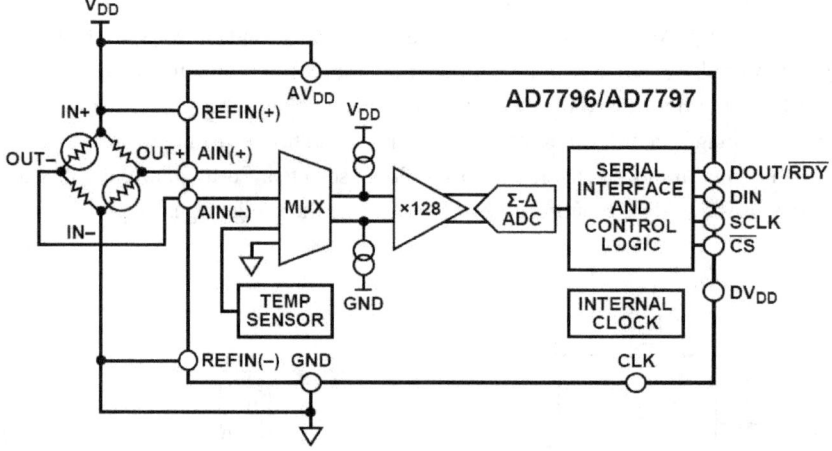

Fig. 3.60: Low Power, 16-/24-Bit Sigma-Delta ADC for Bridge Sensors[13]

$$D_{OUT} = 2^{24} \cdot 128 \cdot \frac{AIN}{V_{REF}} \qquad [77]$$

Fig. 3.60 shows the circuit diagram of an ADC for the direct readout of measuring bridges. The 24-bit AD-converter AD7797 has a differential input, which is optimized for reading out measuring bridges. For a reference voltage of $V_{REF} = 3.3$ V, a measuring range of 25 mV occurs with a voltage resolution of $V_{LSB} = 1.5$ nV.

---

[13] Datasheet AD7796, Analog Devices, 2006

Parameters of Pressure Sensors

In simple pressure sensors, only the four piezo-resistive resistors are integrated, which are connected as a bridge as shown in Fig. 3.56. These sensors have a ratiometric output signal. Tab. 3.9 shows typical data of the three pressure sensor types.

	AT2 (Epcos)	AK2 C27 (Epcos)	HDU M100 D (First Sensor)
Type	Absolute Pressure	Gage Pressure (Barometric)	Differential Pressure
Measuring Range	1.6 bar	0.1 bar	0.1 bar
Temp. compensated	no	no	no
Calibrated	no	no	no
Max. Overpressure	4 bar	0.5 bar	2 bar
Sensitivity	45 ... 95 mV/bar	350 ... 700 mV/bar	300 ... 1200 mV/bar
Temperature Coefficient of Sensitivity	$\alpha = -2.5 ... -1.9 \cdot 10^{-3}$ /K $\beta = +3.0 ... +8.0 \cdot 10^{-6}$ /K^2	$\alpha = -2.4 ... -2.0 \cdot 10^{-3}$/K $\beta = +3.0 ... +8.0 \cdot 10^{-6}$ /K^2	$\alpha = -3.5 \cdot 10^{-3}$ /K
Non-Linearity	± 0.3 % FS	± 1 % FS	± 1.1 % FS
Zero Point Offset	± 30 mV at p = 0 bar	± 25 mV at Δp = 0 bar	± 40 mV at Δp = 0 bar
Temperature Coefficient of Offset	± 10 µV/V/K	± 4 µV/V/K	- 36 µV/V/K
Interfaces	analogous	analogous	analogous
Supply Voltage	0 ... 10 V	0 ... 10 V	0 ... 12 V
Bridge Resistance	2.6 ... 4 kΩ	2.6 ... 4 kΩ	2.8 ... 3.8 kΩ
Package	TO39	DIL	DIL

Tab. 3.9: Parameters of piezo-resistive pressure sensors (all Date at V_{DD} = 5 V)

Pressure sensors with integrated evaluation electronics

In addition to the simple pressure sensors, which only contain the four piezoresistive bridge resistors, there are also sensors with integrated evaluation electronics. This takes over, parallel to the control and measurement of the sensor elements, the signal processing (e.g., temperature compensation), as well as the digital interface function.

In addition to the piezoresistive evaluation method, capacitive evaluation is also possible with fully integrated pressure sensors. Basically, the capacitive evaluation method offers lower temperature dependence and better long-term stability compared to the piezoresistive method. However, in a direct comparison of the two technologies using commercially available sensors, the differences are only marginal.

	LPS22HB [14]	ICP 101 [15]
Evaluation Method	Piezoresistive	Capacitive
Type	Absolute Pressure	Absolute Pressure
Measuring Range	260 ... 1260 hPa	300 ... 1100 hPa
Resolution	0.02 Pa (24 bit)	0.01 Pa (24 bit)
Max. Overpressure	6000 hPa	6000 hPa
Total Measuring Error	± 50 Pa (-20 °C ... 80 °C)	± 100 Pa (0 °C ... 65 °C)
Noise	0.6 Pa (rms)	0.4 Pa (rms)
Interfaces	Digital (SPI, I^{2}C, I3C)	Digital (I^{2}C)
Supply Voltage	1.7 ... 3.6 V	1.8 V ± 5 %
Power Consumption	12 µA	10 µA
Package	2 x 2 x 0.73 mm^3	2 x 2 x 0.72 mm^3
Price (5000 pieces) as of 2020	1.08 €	1.31 €

Tab. 3.10: Data comparison of a piezoresistive and a capacitive pressure sensor

Fig. 3.61 shows a capacitive pressure sensor. The top view shows a round sensor structure in the center of the silicon chip, which is suspended from two posts. The sensor structure consists of two membrane layers with a hermetically sealed space between them. If the pressure to be measured is greater than the reference pressure in the interspace, the membrane bulges inwards and the capacitance increases. In the reverse case, the diaphragm bulges outward and the capacitance decreases. To reduce production- or temperature-related stress influences on the sensor membranes, the entire capacitive structure is suspended from two posts. The posts also provide the electrical contact between the sensor plates and the evaluation electronics located underneath the structure.

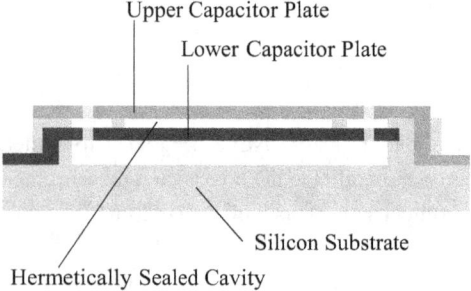

Fig. 3.61: Schematic structure of a capacitive pressure sensor

[14] Datasheet LPS22HB, STMicroelectronics, 2019
[15] Datasheet ICP101, TDK InvenSense, 2016 - 2019

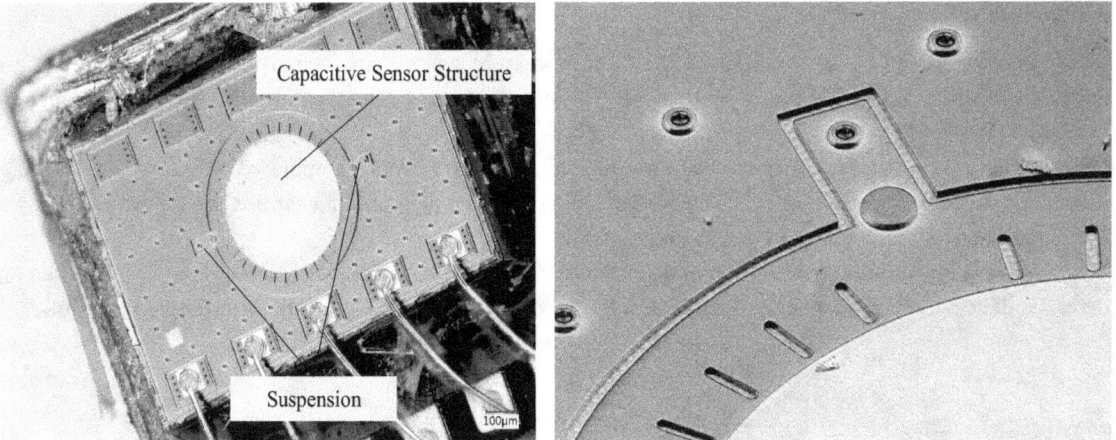

Fig. 3.62: Capacitive pressure sensor (ICP101 TDK InvenSense)

Applications of Pressure Sensors

If one classifies pressure sensors according to their measuring principle, they can be divided into three types of sensors: absolute, differential and gage pressure sensors. All have the action of the measuring pressure on a membrane in common. The differences are in the respective reference pressure.

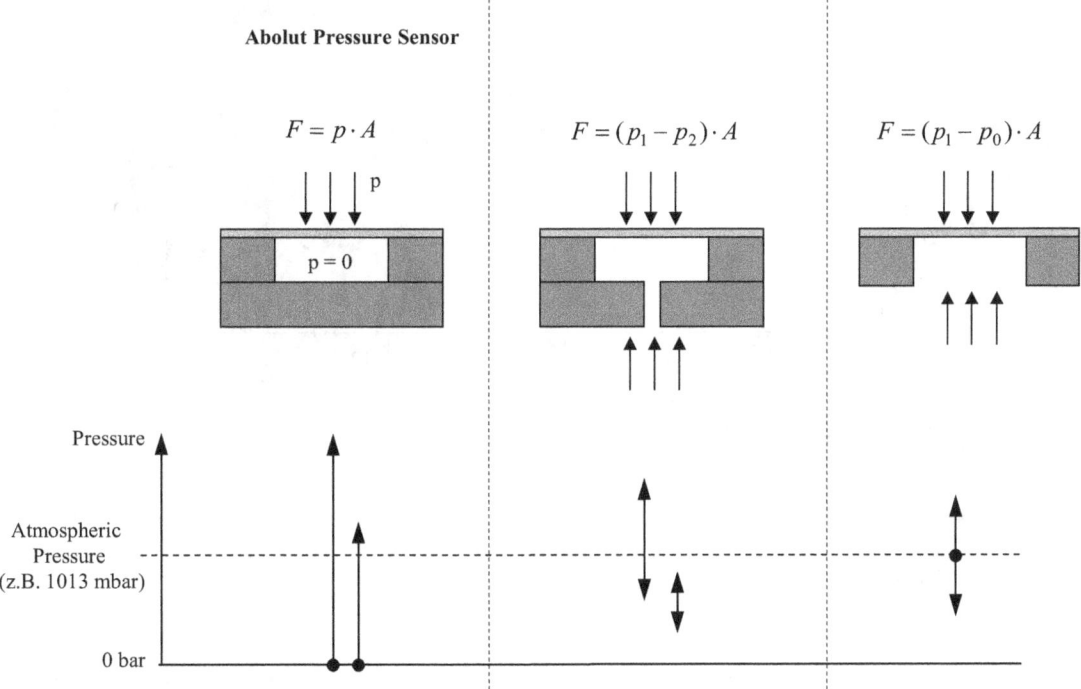

Fig. 3.63: Overview of the different pressure sensor types and their measuring ranges

Absolute Pressure Sensor

Absolute pressure sensors measure an external pressure against an internal reference pressure. This reference pressure is provided by a hermetically sealed cavity, and can be adjusted during the production process. Very often, a reference pressure close to zero is selected, so that the measured pressure difference on the membrane corresponds to the outer absolute pressure. A typical example of the absolute pressure measurement are tire pressure sensors. They measure the internal pressure of a car tire and report wirelessly a drop in pressure at the respective control electronics.

Another application example is height determination with the help of the absolute pressure. It is based on the dependence of the atmospheric pressure at the level of the measurement location relative to sea level.

$$\frac{p}{p_0} = \left(1 - \frac{h}{44330 \ m}\right)^{5,255} \quad [78]$$

with p - *Measured pressure*
p_0 - *Atmospheric pressure on sea level*
h - *Altitude above sea level in [m]*

With the pressure function described in Eq. [78], a height sensitivity of 0.12 hPa/m is obtained at sea level. Due to the strong influence of weather conditions on atmospheric pressure (980 ... 1040 mbar), this measurement is only suitable for measuring relative changes in altitude. *Fig.* 3.64 shows the example of a pressure measurement on several floors of the building of the University of Applied Sciences Bielefeld. For this purpose, all floors (basement floors (U1-U3) and upper floors (O4-O2)) of the building were approached by elevator and the pressure values occurring were recorded. For this measurement, the absolute pressure sensor LPS25HB[16] with no further noise cancelation was used. Despite the relatively high noise level of the signal, the measured values in *Fig.* 3.64 can be clearly assigned to the individual floors.

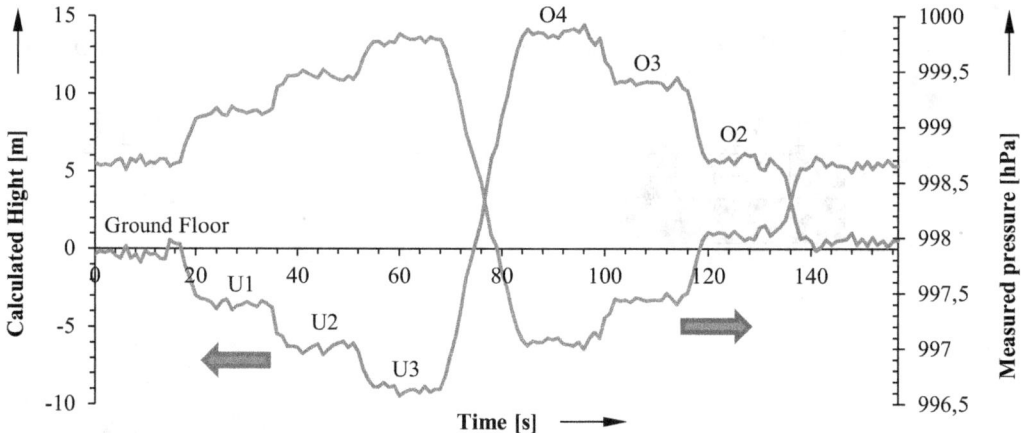

Fig. 3.64: Pressure and height measurement in the building of the FH-Bielefeld

---

[16] Datasheet LPS25HB, STMicroelectronics, 2016

Differential Pressure Sensors

Differential pressure sensors measure the pressure difference between two measuring media. A typical application example is flow measurement. For this measurement, an orifice is introduced into a flowing gas, in which it dams the flow. One measures, in this case, a higher pressure before the orifice than downstream of the orifice. Based on the pressure difference the flow velocity can be determined.

$$\theta = \frac{dV}{dt} = \alpha \cdot \varepsilon \cdot A\sqrt{2\Delta p/\rho} \qquad [79]$$

with θ - Flow rate
 A - Area of the orifice opening
 ρ - Specific density of the flowing medium
 Δp - Measured pressure difference
 $\alpha = 0.6 - 0.8, \; \varepsilon = 0.9 - 1$

Fig. 3.65: Differential pressure sensor (Source: First Sensor AG)

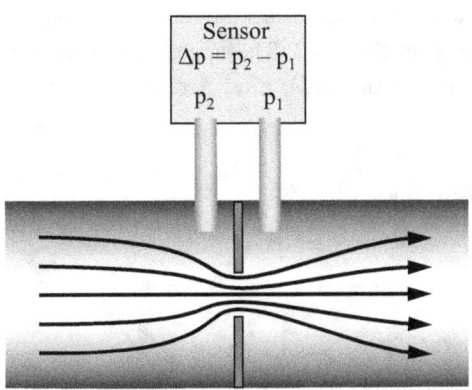

Fig. 3.66: Measuring setup for the determination of the flow velocity by using a differential pressure sensor

Gage Pressure Sensors

Gage pressure sensors are actually differential pressure sensors, in which the reference pressure is represented by the external atmospheric pressure. Typical examples of such a measurement are level measurement of liquids. Here the pressure into the liquid is equal to the sum of the atmospheric pressure and the pressure, which is caused by the weight of the liquid column. In Eq. [80] it is seen, that for the determination of the filling height, the pressure difference between liquid and atmosphere is used.

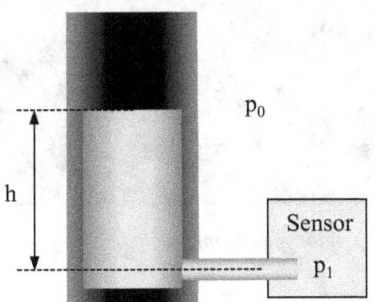

$$h = \frac{p_1 - p_0}{\rho \cdot g} \qquad [80]$$

with h - Filling height
 $p_1 - p_0$ - Pressure difference between atmosphere and liquid
 ρ - Specific density of the liquid
 g - Gravity

Fig. 3.67: Determination of the filling height by use of a gage pressure sensor

Microphones

Microphones are a special class and are commonly not considered to be pressure sensors. They are based on the dynamic measurement of air pressure and are thus similar to pressure sensors. In recent years, there was a rapid development of MEMS microphones in this area. Driven by mobile technology, the electret microphones used previously were almost completely replaced and thus the size and the cost of microphones for speech transmission has been significantly reduced.

Fig. 3.68 shows the structure of a MEMS microphone with capacitive evaluation. A double membrane is stretched over a cavity. The top membrane is heavily pierced and serves as a rigid counter electrode. If a sound signal arrives at the microphone, the sound pressure produces a deflection of the lower membrane, and increases the distance to the counter electrode, whereby the capacitance between the membrane and the counter electrode is reduced.

The cavity of MEMS microphones is not hermetically sealed, so that there is atmospheric pressure inside, and slow pressure changes do not affect the microphone output signal. This property is responsible for the lower cut-off frequency in Fig. 3.72. The upper frequency limit is determined by the natural resonance of the membrane structure and leads, after a resonance rise, to a strong decrease in the sensitivity at higher frequencies.

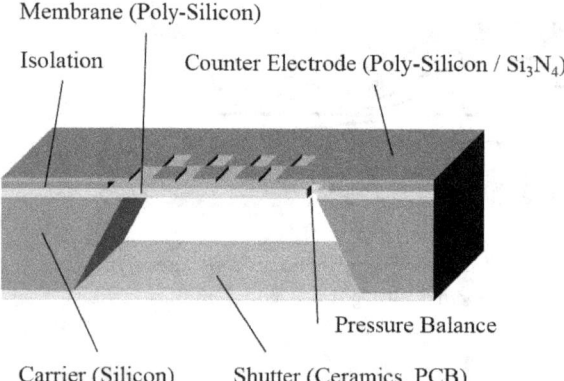

Fig. 3.68: Structure of a capacitive MEMS microphone

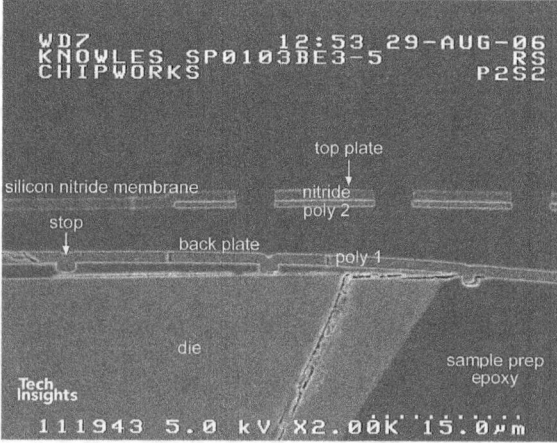

Fig. 3.69: Chip photo of a capacitive MEMS microphone and cross section through the membrane
(Image source on the right: Chipworks – Tech Insights)

Recent developments in MEMS microphones show the trend to use a differential capacitance principle. Here, two back plates in the form of perforated free-floating layers surround the moving diaphragm. As already discussed for the capacitive evaluation system of accelerometers, the differential capacitance principle (see Eq.[25]) leads to a higher sensitivity and a higher linearity of the sensor characteristic. In the case of microphones, the higher linearity results in a lower distortion factor in the audio transmission, and with the higher sensitivity, a better signal-to-noise ratio (SNR) of the microphone.

Fig. 3.70 shows the structure of a MEMS microphone based on the differential capacitance principle. The center electrode is a closed membrane, which is deflected by the sound waves. The two counter electrodes are designed as a grid membrane and serve as fixed electrodes.

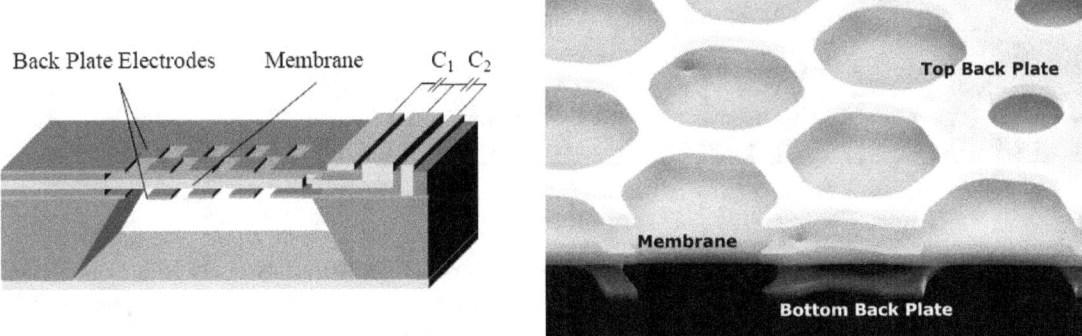

Fig. 3.70: Structure of a MEMS microphone with a differential capacitance membrane
(Right image source: Infineon Technologies AG)

Parameters of MEMS Microphones

The parameters of a microphone are divided into input and output parameters. Both parameters are specified in decibels (dB). The input parameters are referenced to the sound pressure of the human hearing threshold (0 dB = 20 µPa) and are specified in dB(A). The output parameters for microphones with analog output are referenced to 1 V (the amplitude of the sinusoidal signal) and measured in dBV.

The sensitivity of a microphone is determined at an acoustic input reference level of 94 dB (SPL). It corresponds to a sound pressure of 1 Pa. In addition to sensitivity, signal-to-noise ratio (SNR) plays a key role in assessing the quality of a microphone. Subtracting the SNR from the sensitivity gives the average noise level of the microphone at the output.

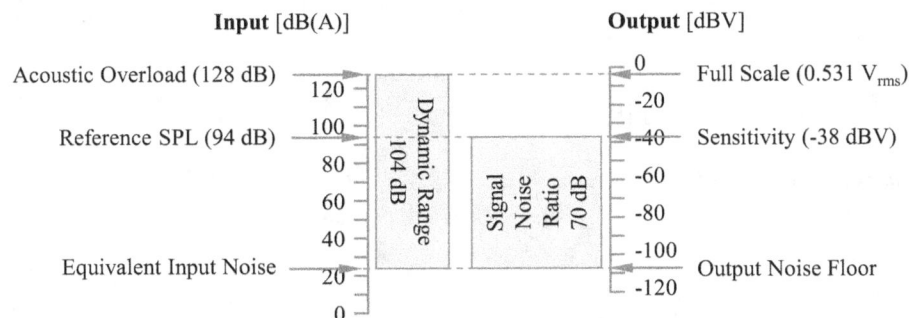

Fig. 3.71: Input and output parameters of the microphone ICS-40800

Parameters	Typical Values
Analog Output FS	0.531 V_{rms} (for 128 dB SPL)
Sensitivity (1 kHz, 94 dB SPL)	- 38 dBV
Signal to Noise Ratio (SNR)	70 dB
Dynamic Range	104 dB
Noise Floor	- 108 dBV
Lower Cutoff Frequency (- 3 dB)	80 Hz
Upper Cutoff Frequency (+ 5 dB)	> 20 kHz
Power Supply	1.6 V ... 3.6 V
Power Consumption	155 µA
Package	LGA bottom port 4 x 3 x 1.2 mm³

Tab. 3.11: Selected data from the microphone ICS-40800 [17]

Since MEMS microphones are designed for use in smartphones, their frequency range of these is adapted to the frequency spectrum of the human voice. To achieve this, a lower cutoff frequency of approx. 100 Hz and an upper cutoff frequency of approx. 10 kHz are sufficient. The upper cutoff frequency is determined by the resonant frequency of the microphone's diaphragm; while the lower one is determined by the dimensions of the pressure balance hole (see Fig. 3.68).

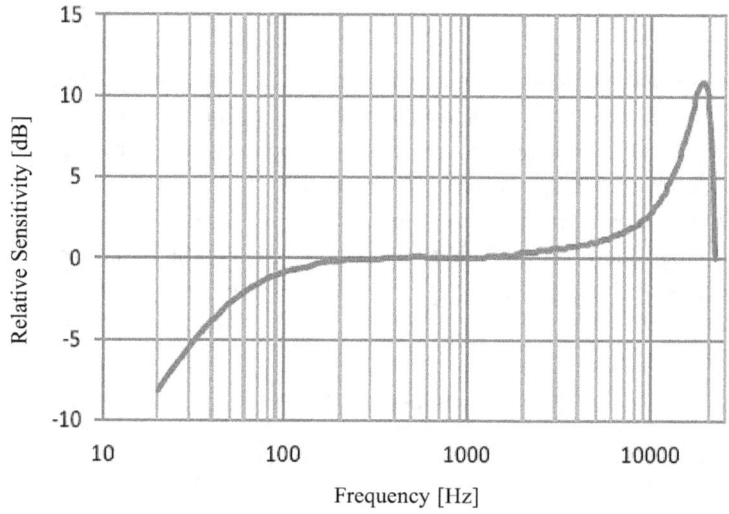

Fig. 3.72: Typical frequency response of a MEMS microphone

---

[17] Datasheet ICS-40800, TDK InvenSense Inc, 2020

4 Actuators

MEMS actuators are much more difficult to realize compared to MEMS sensors. This is mainly due to the very low forces and deflections that a MEMS system can generate. Therefore, MEMS actuators are mainly used to drive motions in MEMS systems (e.g., gyroscopes or oscillators). However, they are also used in optical (e.g., mirror systems), fluidic (e.g., inkjet heads) or acoustic systems (micro loudspeakers).

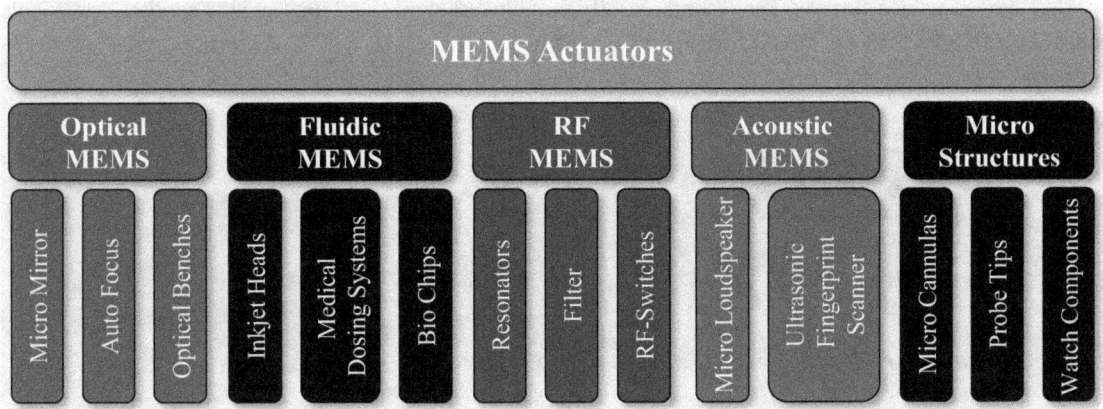

Fig. 4.1:   Overview of available MEMS actuators

Primarily electrostatic and piezoelectric forces are used as drive forces. They have the advantage that they can be operated with relatively low currents and are easy to integrate.

For the excited movements in the microactuator, only very small deflections can be achieved due to the dimensions of the microsystem. Furthermore, continuous rotary motions, as seen in Fig. 4.2, are unfavorable when used as microactuators. This unfavorability is mainly due to the presence of frictional contacts in this arrangement. Hence, continuous rotary motions should be avoided for two reasons. First, frictional contacts generate dust particles that can clog the small cavities in the microsystem and lead to system failure. Second, adhesion effects (sticking effects) at contact points produce unpredictable hysteresis properties in the characteristic curve of the microsystems.

For this reason, freely swinging designs are found in almost all microactuators, which are either statically deflected or in which oscillation is induced.

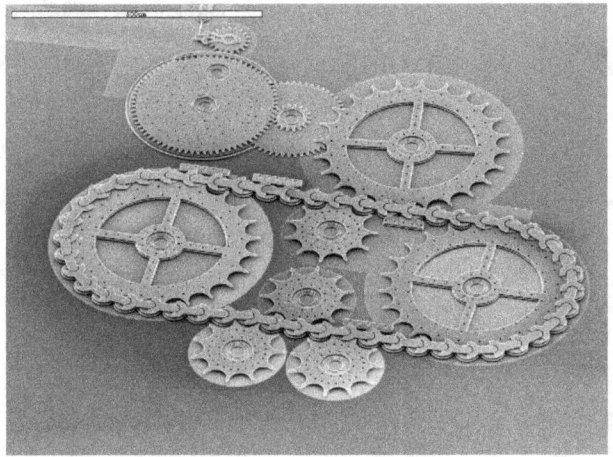

Fig. 4.2:   Technology demonstrator of a chain transmission using MEMS technology

Source: Sandia National Laboratories
(www.sandia.gov/media/NewsRel/NR2002/chain.htm)

4.1 Types of Drives

Electrostatic Excitation

Putting a voltage on the capacitor plates, an electrostatic force acts on the electrodes of the capacitor. The force is proportional to the surface area of the capacitor (A) and indirectly proportional to the square of the distance between the electrodes (d). Furthermore, the force is proportional to the square of the applied voltage. Eq. [82] shows this dependency. The technologically feasible distances (1 - 2 μm) and the maximum available voltages (1.8 - 5 V) limit the forces, producible in MEMS sensors.

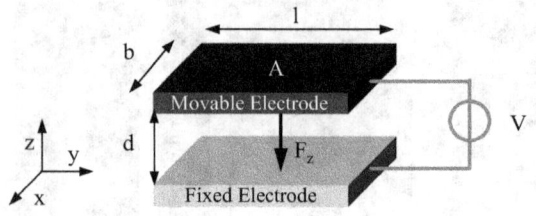

Fig. 4.3: Electrostatic force on a plate capacitor in the z-direction

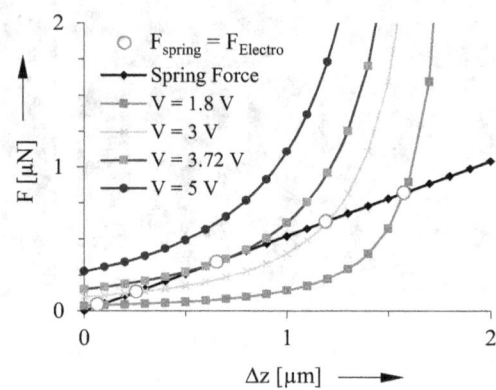

Fig. 4.4: Electrostatic forces on a plate capacitor arrangement ($A = 100 \cdot 100$ μm²; $d = 2$ μm) and the counterforce on a silicon beam spring ($100 \cdot 2 \cdot 2$ μm³)

$$W = \frac{C \cdot U^2}{2} \qquad [81]$$

$$F_z = \frac{dW}{dz} = \frac{1}{2}\varepsilon_0 \cdot U^2 \frac{l \cdot b}{d^2} \qquad [82]$$

The electrostatic force counteracts the spring force of the silicon spring, which rises linearly with the deflection (Δz). In the equilibrium case, the electrostatic force is equal to the spring force. The equilibrium points are shown in Fig. 4.4. These are the intersections of the two graphs (electrostatic force and spring force). In addition to the strong non-linearity of the electrostatic force, it can be seen in Fig. 4.4, that for voltages less than 3.72 V, two intersection points exist. For voltages above 3.72 V no intersection exists. This means that for voltage larger than 3.72 V the arrangement shows no state of equilibrium, and the plates are attracted until the mechanical contact with the counter occurs. This voltage is named the snap-in voltage and can be calculated by Eq. [83].

$$U_{SnapIn} = \sqrt{\frac{8k \cdot d^3}{27\varepsilon_0 \cdot A}} \qquad [83]$$

with k - Spring constant
     A - Area of the plate capacitor
     d - Initial distance of the electrodes
     ε_0 - Permittivity of air

The non-symmetry and the snap-in effect are undesirable characteristics, which greatly restricts the use of this arrangement. In practice, it can only be used in a closed-loop arrangement in which the excitation oscillation is controlled by a parallel capacitance measurement. The closed loop arrangement gives the advantage of the relatively low necessary excitation voltages, and a stable vibration exciter.

Another possibility to generate an electrostatic excitation is to use the horizontal force. It occurs in the illustrated arrangement (Fig. 4.5), wherein two fixed electrodes surround the movable electrode. The attraction forces in the z-direction cancel each other out in this arrangement and only the force in the y-direction remains, after Eq. [84].

$$F_y = \frac{dW}{dy} = \frac{1}{2} \cdot U^2 \cdot 2 \cdot \varepsilon_0 \frac{b}{d} \qquad [84]$$

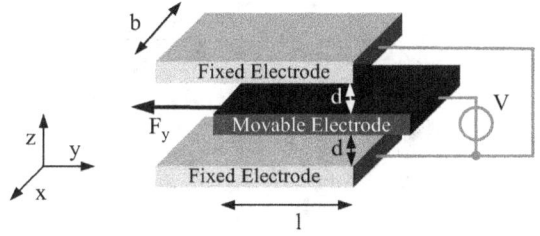

Fig. 4.5: Electrostatic force on a double-plate capacitor in y- direction

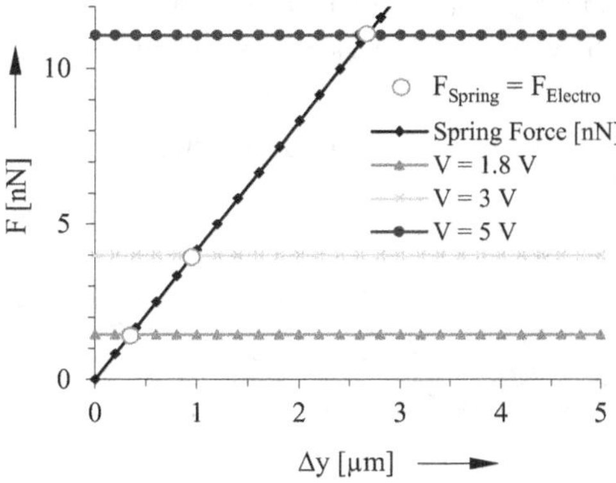

Fig. 4.6: Electrostatic force on a double-plate capacitor structure ($A = 100 \cdot 100$ μm²; $d = 2$ μm) and the counterforce of a silicon-beam spring ($100 \cdot 2 \cdot 2$ μm³)

For this arrangement, the electrostatic attraction in the y-direction is independent of the deflection, which thus allows the realization of relatively large displacements in small gaps. One disadvantage is the significantly lower forces that can be achieved. While the forces in the arrangement in Fig. 4.4 were still in the μN-range, here, they are only in the nN-range. Furthermore, forces also occur in the z-direction in the arrangement in Fig. 4.5. Ideally, they compensate each other. However, if the gap distances are not completely symmetrical, a resulting attraction force of the moving electrode to one of the fixed electrodes occurs. To prevent this from leading to undesirable tilting movements in the capacitor gap, the spring suspension must be designed to be sufficiently stiff in the z-direction.

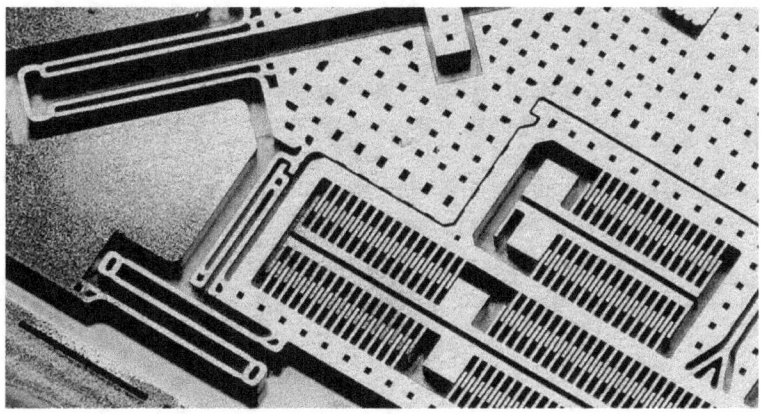

Fig. 4.7: Finger structure for the capacitive excitation of a gyroscope

Piezoelectric Excitation

The piezoelectric effect describes the generation of an electrical voltage on a solid when it is elastically deformed. Conversely, a mechanical deformation also occurs on the piezo material when a voltage is applied, which thus induces an electric field in the material. This effect is called the inverse piezoelectric effect. Both the piezoelectric effect and the inverse piezoelectric effect are only found in materials that have permanent dipoles or form dipoles during elastic deformation.

Piezoelectric actuators are based on the inverse piezoelectric effect. They can be operated either quasistatically or resonantly. Furthermore, in addition to the longitudinal effect shown in Fig. 4.8, the transversal effect can also be used. In the case of the longitudinal effect, the change in length and the electric field are parallel to each other. In the case of the transversal effect, they are perpendicular to each other. In addition, mechanical shear stresses can be generated in some materials.

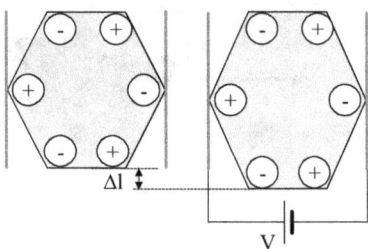

Fig. 4.8: Inverse piezoelectric effect

The inverse piezo effect can be described as a linear relationship between the applied electric field strength and the change in length.

$$\varepsilon = d \cdot E \qquad [85]$$

with ε - Relative change in length
      d - Piezoelectric coefficient
      E - Electric field strength

The relative length change described in equation [85] applies to the mechanically unloaded case. This means that no mechanical force is exerted on the piezo element. In reality, however, a counterforce is usually generated when the piezo actuator is deflected. This can be, for instance, the force at the deflection of a diaphragm or simply the weight force of a mass. In this case, the equation [85] is extended by a modulus, which describes the linear elastic deformation of the material.

$$\varepsilon = d \cdot E + s^E \cdot \delta \qquad [86]$$

with s^E - Inverse modulus of elasticity
      δ - Mechanical stress, force per area

In equation [86] shows that the electric field strength is a three-dimensional vector with the three spatial directions (E_X, E_Y, E_Z). For the mechanical stress (δ), in addition to the three spatial directions (δ_x, δ_y, δ_z) the shear stresses (τ_{xy}, τ_{xz}, τ_{yz}) are added, so that the stress vector is six-dimensional. As a result, a six-dimensional result vector is also obtained, which, in addition to the relative length changes (ε_x, ε_y, ε_z) also contains the three shear angles (γ_{xy}, γ_{xz}, γ_{yz}).

To describe the material properties, a 3x6 matrix is used for the piezoelectric constant, and a 6x6 matrix is used for the inverse elastic modulus. The followed expression is thus obtained to give a general description of the inverse piezoelectric effect:

$$\varepsilon_\mu = \sum_{i=1}^{3} d_{i\mu} E_i + \sum_{k=1}^{6} s^E_{\mu k} \delta_k \qquad [87]$$

For the mechanically unloaded case, no opposing forces hinder the expansion of the piezoelectric material, and thus, no mechanical stress occurs. The second summand of Eq. [87] becomes zero and we obtain Eq. [88], which reproduced here in matrix notation:

$$\begin{pmatrix} \varepsilon_x \\ \varepsilon_y \\ \varepsilon_z \\ \gamma_x \\ \gamma_y \\ \gamma_z \end{pmatrix} = \begin{pmatrix} d_{11} & d_{21} & d_{31} \\ d_{12} & d_{22} & d_{32} \\ d_{13} & d_{23} & d_{33} \\ d_{14} & d_{24} & d_{34} \\ d_{15} & d_{25} & d_{35} \\ d_{16} & d_{26} & d_{36} \end{pmatrix} \begin{pmatrix} E_x \\ E_y \\ E_z \end{pmatrix} \qquad [88]$$

When describing the piezoelectric properties, the material can usually be characterized by a few coefficients since many coefficients are zero and others are equal due to symmetry constraints. For example, the matrix of quartz can be characterized by only two coefficients. If the coefficient matrix of quartz (Eq. [89]) still consists of five nonzero entries, the number of independent coefficients is reduced to two by symmetry constraints with $d_{11} = -d_{12} = -d_{26}/2$ and $d_{14} = -d_{25}$.

$$\begin{pmatrix} \varepsilon_x \\ \varepsilon_y \\ \varepsilon_z \\ \gamma_x \\ \gamma_y \\ \gamma_z \end{pmatrix} = \begin{pmatrix} d_{11} & 0 & 0 \\ d_{12} & 0 & 0 \\ 0 & 0 & 0 \\ d_{14} & 0 & 0 \\ 0 & d_{25} & 0 \\ 0 & d_{26} & 0 \end{pmatrix} \begin{pmatrix} E_x \\ E_y \\ E_z \end{pmatrix} \qquad [89]$$

Modified lead zirconate titanates (PZT) are generally used as piezoelectric materials in microsystems technology, while lead magnesium niobates (PMN) are generally used in low-voltage actuators. Both materials are ceramics that can be cut from a block or applied as a layer. For flexible applications, piezoelectric foils are available to the user, such as polyvinylidene fluoride foils.

MEMS

Material	Piezoelectric Coefficients d [10^{-12} m/V]	Curie temperature T [°C]
Quartz	$d_{11} = -d_{12} = -d_{26}/2 = 2.31$ $d_{14} = -d_{25} = 0.73$	550
PZT Lead Zirconate Titanate	$d_{15} = 265 \ldots 765$ $d_{31} = -60 \ldots 270$ $d_{33} = 380 \ldots 590$	190 … 360
PMN Lead Magnesium Niobate	$d_{31} = -600 \ldots -800$ $d_{33} = 1 \ldots 2$	80 … 110
PVDF Polyvinylidene Fluoride	$d_{31} = 18 \ldots 28$ $d_{32} = 0.9 \ldots 4$ $d_{33} = -20 \ldots -35$	170 … 200

Tab. 4.1: Properties of piezoelectric materials

Since the linear expansion of piezoelectric materials in the quasi-static case is relatively small (max. 0.1 % of the actuator length), resonance peaks or lever effects are used to amplify the effect. A good example of the use of the lever effect are piezo buzzers, which consist of a layer stack with a PZT layer and a brass sheet. If a voltage is applied to the PZT material, the diameter of the PZT disc increases (transversal effect) and bulges the entire layer stack by a multiple of the original linear expansion of the PZT material.

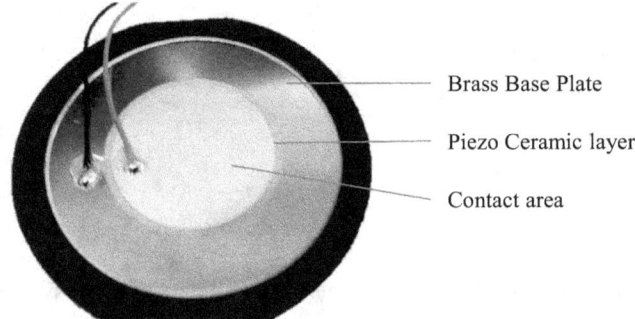

Fig. 4.9: Piezo buzzer for generating signal tones (1 kHz)

4.2 Optical MEMS

Micro-Opto-Electromechanical Systems (MOEMS) are microactuators that influence the beam path of light sources. This can be done to generate an image (e.g., overhead projectors), to analyze materials and surfaces (e.g., laser scanning microscopy) or to influence signal paths in optical data transmission (e.g., shutters or filters).

For the fabrication of MOEMS, the same processes are used, as in semiconductor and MEMS technology. Optically active layers, which also complement these processes, are in most cases highly reflective layers that can be produced with a very low level of unevenness and roughness by special mechanical and chemical processes. Aluminum or gold layers are often used as mirror layer materials for this purpose.

Micromirrors

At just under a billion dollars a year, micromirrors are the highest-selling MOEMS application. They are divided into digital and analog systems. For analog working micromirrors, the tilt angle of the mirror can be continuously adjusted. This allows an image to be projected line by line, similar to a CRT screen. Such mirrors are mainly used in laser projectors or laser scanners.

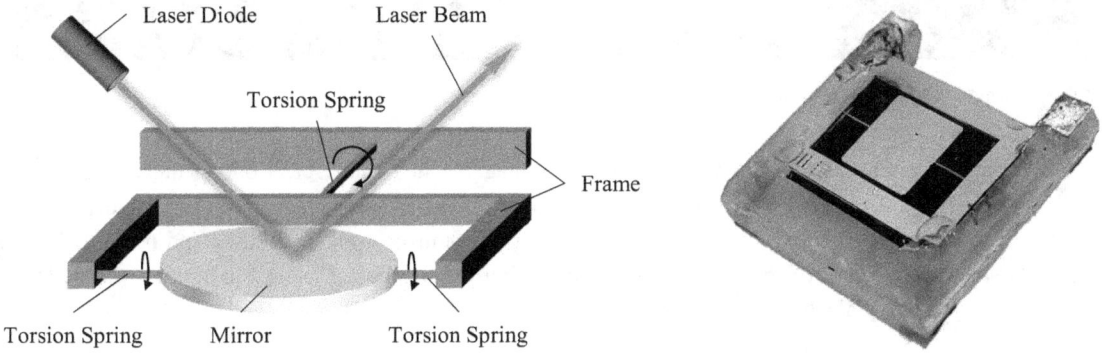

Fig. 4.10: Schematic structure of a micromirror and a chip photo of a mirror element

As a rule, digitally operating micromirror arrays use an array of mirrors, which can switch the pixels on or off by tilting. Each mirror corresponds to a pixel. For high-resolution image displays (4k-UHD 3840 x 2160), this results in a mirror count of 8 million mirrors per device. Due to their small size (approx. 10 x 10 µm^2) and mass, the mirrors can be moved very quickly. This property is used to generate brightness modulations of the individual pixels. Fig. 4.11 shows the image generation on DLP arrays.

Projectors for the mass market are 1-chip devices. They contain a DLP chip, a color separator in the form of a color wheel and a white light source. A high-pressure mercury vapor lamp or a white power LED can realize the light source. The latter is much less expensive and longer-lasting, but has a lower light output. The color wheel contains the RGB color filters, which are passed through in sequence by rotating the wheel. In the simplest application, this wheel rotates at 3000 revolutions per minute. More sophisticated projectors use color wheels with 6-color segments and rotation speeds of 6000 rpm. This prevents the "rainbow effect" on bright vertical picture edges, which occurs with relatively slow color changes. For small portable projectors, 3-LED systems are often used. They contain a red, a green and a blue LED as light source, and thus replace the relatively complex color separation by means of a color wheel.

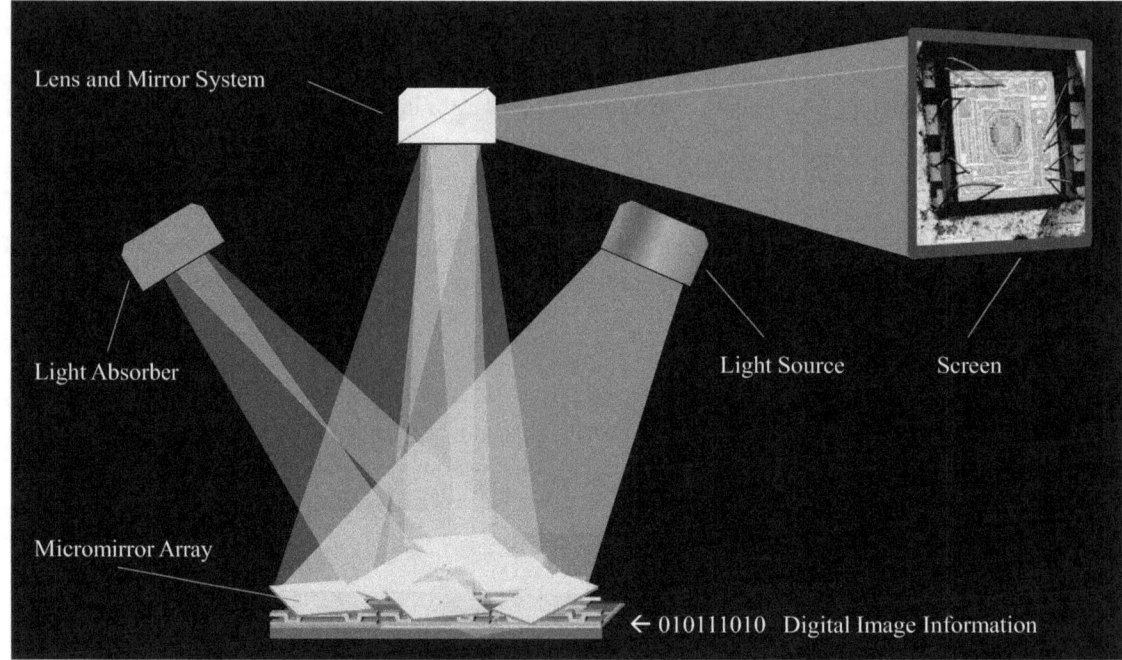

Fig. 4.11: Image creation in digital light processing (DLP)

Each mirror can be tilted by ± 12° by applying a voltage to the counter electrode of the yoke plate. The respective mirror deflects until it comes to a mechanical stop. This ensures an exact tilt angle independent of temperature, material or voltage variations. Fig. 4.13 shows a mirror array with tilted and non-tilted mirror elements. The substructure of the mirror system, including the mechanical stop for limiting the deflection angle, can be seen very nicely in this image.

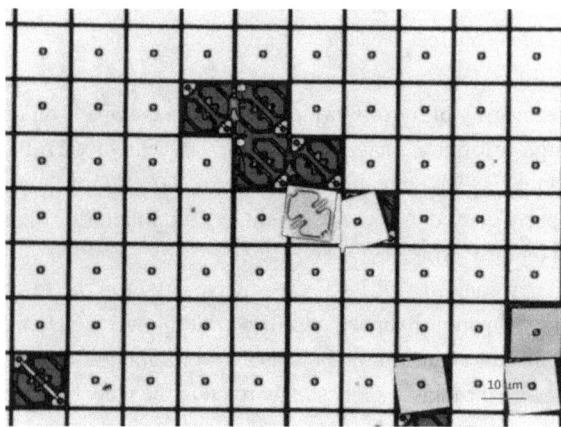

Fig. 4.12: Micromirror array with defective mirror elements

Fig. 4.13: Detailed view with tilted mirror elements (Source: Courtesy Texas Instruments)

Technology for the Realization of Micromirrors

Fig. 4.14 shows the simplified structure of a micromirror array. It contains the mirror elements, which can be tilted via a torsion spring. The tilting is controlled by the electrostatic attraction between the yoke plate and the counter electrode integrated in the silicon substrate.

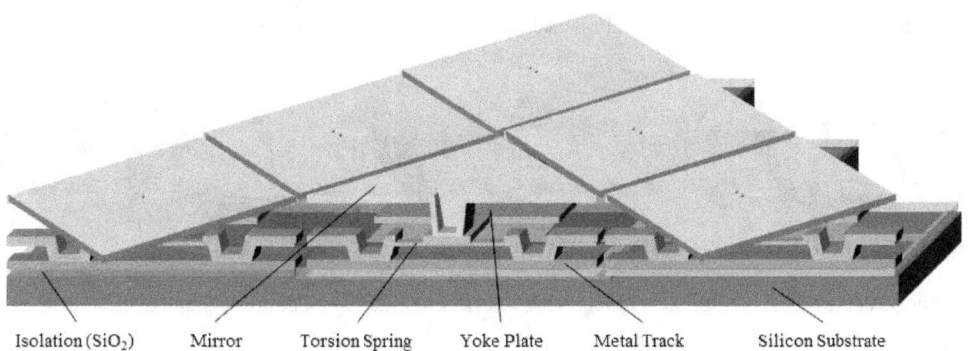

Fig. 4.14: Illustration of the layer structure on a mirror array

The micromirror arrays are fabricated using a sacrificial layer technology. The individual steps are shown in Fig. 4.15. The starting point is a silicon substrate, which contains the drive electronics and the counter electrodes of the yoke plate actuator. The micro-mirror structure is built on top of this pre-prepared substrate. In order not to transfer the unevenness of the layer systems of the electronics to the mirror surface, the substrate is ground flat by means of a chemical mechanical polishing (CMP) process before the mirror structure is built up. In addition, the selection of possible technologies is limited to low-temperature process steps (T < 400 °C) in order not to destroy the already existing evaluation electronics.

After applying and structuring the conduction tracks, used to control the micromirrors, the first sacrificial layer is applied. A photoresist is used as the sacrificial layer, which is cured by UV radiation during photolithographic structuring. The torsion spring and the yoke plate are built up on this layer. The second sacrificial layer specifies the distance between the yoke plate and the micromirror and consists of the same UV-cured photoresist as the first sacrificial layer. As the last functional element, the mirror elements are implemented on the entire layer stack. Before the sacrificial layers are removed using the plasma etching process, thus exposing the micromirrors, chip separation is prepared. Chip separation is necessary because the micromirror arrays are realized as a batch process on an 8-inch wafer. For the separation, the silicon substrate is sawed in order to separate later the chips by breaking at these predefined points. The relatively complex separation process (sawing and subsequent breaking) is necessary because, during the sawing process, there is strong contamination with particles (sawdust) and cooling liquids, which would destroy the freestanding micromirrors or clog their fine gaps.

Silicon Substrate with
CMOS Evaluation Circuit
and Counter Electrode

SiO_2-Layer
→ Oxidation
→ Chemical Mechanical
    Polishing Process (CMP)

Metal Conduction Track
→ Sputtering
→ Photolithographic Structuring

Sacrificial Layer 1 (Photoresist)
→ Spin ON
→ Photolithographic Structuring

Torsion Spring (Al-Alloy 60 nm)
→ Sputtering
→ Photolithographic Structuring

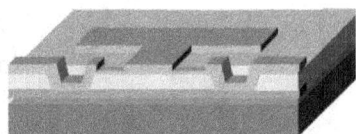

Yoke Plate (Al-Alloy)
→ Sputtering
→ Photolithographic Structuring

Sacrificial Layer 2 (Photoresist)
→ Spin ON
→ Photolithographic Structuring

Mirror (Al-Alloy)
→ Sputtering
→ Photolithographic Structuring

Removing
Sacrificial Layer 1 & 2
→ Plasma etching photoresist

Fig. 4.15: Technology steps for the realization of a micromirror

4.3 Fluidic MEMS

Fluidic MEMS actuators are often used for dispensing very small quantities of liquids. In addition to the classic inkjet heads, in which the pump and nozzle system are integrated micromechanically, dispensing systems for medical technology have shown increased visibility in the market in recent years. Here, the low-cost batch production of microsystems is very conducive to the single-use of end-user medical products.

Micropumps

MEMS can be used to transport and dosing liquids. Due to the small dimensions, the quantities of liquid are very limited. Pumps for providing the smallest quantities of liquid create a delivery volume of approx. 100 ml/min at a pressure of one bar. Piezoelectric actuators are usually used as the drive system, either deposited as a thin layer on the MEMS structure or mounted on the MEMS system as a hybrid structure.

Fig. 4.16 shows the micropump produced by Debiotech. It has a dimension of 10 x 6 mm^2 and consists of a glass/silicon/glass stack that is anodically bonded in the wafer composite. The pump is driven by a bonded piezo element. In addition to the active pump elements (pump diaphragm, inlet and outlet valve), the sensor contains a pressure and temperature sensor for function monitoring. The pump has a delivery volume of 2.5 ml/h and is used for individual insulin dosing.

Fig. 4.16: Micropump produced by Debiotech (Source: Debiotech S.A.)

The structure of a MEMS pump is shown in Fig. 4.17. It contains a pump diaphragm and an inlet and outlet valve, which are structured in a silicon wafer by means of etching processes. Furthermore, there is an upper and lower cover, which can also be made of silicon or borosilicate glass. The three wafers are bonded together and afterwards the separated into single chips.

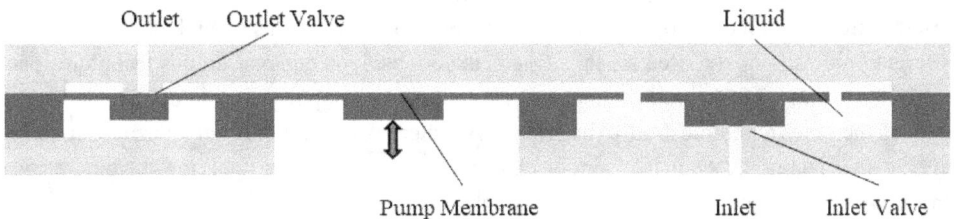

Fig. 4.17: Schematic structure of a micropump

The structure of a MEMS pump is shown in Fig. 4.17. It contains a pump diaphragm and an inlet and outlet valve, which are structured in a silicon wafer by means of etching processes. Furthermore, there is an upper and lower cover, which is made of silicon or borosilicate glass. The three wafers are bonded together, and then the individual chips are separated.

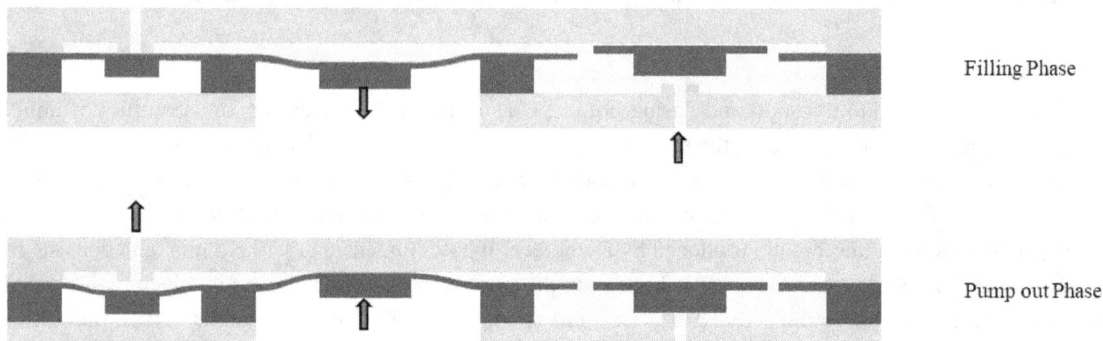

Filling Phase

Pump out Phase

Fig. 4.18: Functional principle of a micropump

Inkjet Printer

One of the first MEMS applications, and to date the top-selling application of fluidic MEMS, are inkjet heads. Inkjet technologies are applied in a wide field. This includes the classic desktop printer, a high-speed printer for personalization of mass printed products up to high-quality poster printers. In addition, there are a number of special applications where functional layers are applied by inkjet technologies. The great advantage of inkjet technology is that a printed image can be produced without having to make printing plates or masks beforehand.

The achievable resolutions of inkjet printing are around 1500 pixels per inch (dpi). It is mainly used for high-quality poster or photo prints. With 600 dpi, commercially-available inkjet printers for office and home use are in the mid-range and are characterized by a very favorable price compared to other desktop printers. Inkjet systems with very low resolutions are used for personalizing printed products (e.g., imprinting address data on offset-printed direct mail). For these applications, high printing speed is particularly important in order to operate the inkjet system in line with offset or gravure systems.

A special application of inkjet technology is the application of functional layers. In electrical engineering, this is used to print the solder mask on circuit boards. However, highly viscous media, such as solder paste used in circuit board assembly, can also be applied using inkjet techniques.

Most inkjet printers are based on the drop-on-demand principle, which generates one ink drop per printed color dot. Two different processes can be used for this. One is the bubble jet technique. In this technique, part of the ink is vaporized by heating. The explosive increase in volume during vaporization forces an ink droplet out of the nozzle, which then flies to the substrate to be printed. Another inkjet technology are piezoelectrically-driven print heads. In this technology, an ink supply chamber is contracted by applying a voltage to a piezoelectric element, which then leads to the ejection of an ink drop.

Actuators

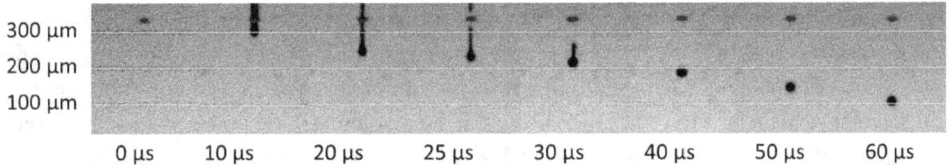

Fig. 4.19: Droplet formation at the inkjet head

The functional elements of a piezoelectrically driven inkjet head (i.e., nozzle, feed channels, pump and valves) are manufactured from silicon using volume-bulk micromechanics. On the one hand, this ensures high manufacturing precision combined with low production costs. On the other hand, silicon is inert to organic solvents and corrosion-free. Fig. 4.20 shows the basic structure of an inkjet head. It consists of the nozzle, the ink channel, a passive inlet valve and the piezoelectric ally-driven membrane pump.

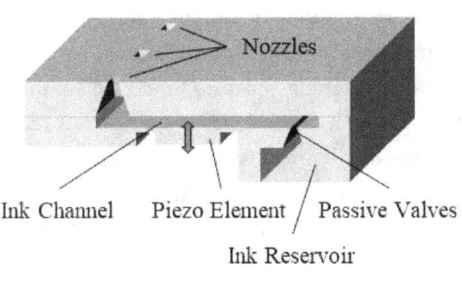
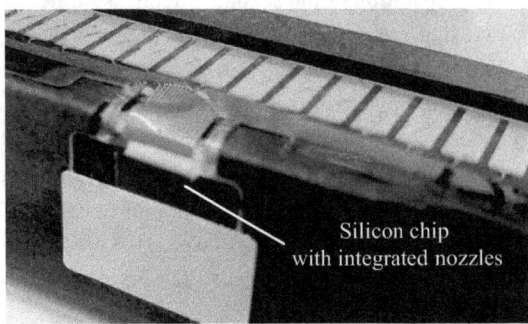

Fig. 4.20: Schematic structure of an inkjet head   *Fig. 4.21: Inkjet head with silicon nozzles*

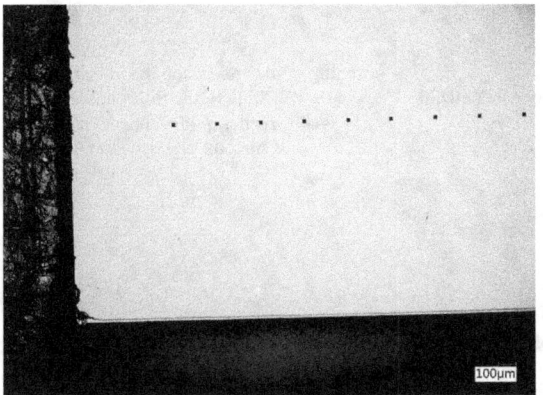

Fig. 4.22: Nozzle outlets within the silicon   *Fig. 4.23: Piezoelectric Elements*

When a voltage is applied to the piezo element, it bends, and the underlying membrane is displaced into the ink channel. Since the passive valve's design strongly impedes the resulting ink flow back into the ink reservoir, the reduced volume resulting from the membrane movement is ejected via the nozzles in the form of a droplet. In the next step, the voltage at the piezo element is reversed. The membrane returns to its original position and the missing ink volume flows back into the ink channel via the valve. The print head is ready for the next ink drop.

Inkjet heads are manufactured using volume micromechanics technologies. In order to etch the buried ink channels into the silicon, two silicon wafers are structured independently of each other and then bonded together. Fig. 4.24 shows a highly simplified technology flow for fabrication. The first silicon wafer contains the nozzle and ink channel structures. They are fabricated using a two-step etching process. To avoid having to perform a photolithographic process for the second etch step on the pre-patterned wafer, it is possible to place the information into a multilayer mask system at the beginning of the process. Here, it is a SiO_2/Si_3N_4 double-layer system. After the first etching step to realize the deep etched nozzle structure, the Si_3N_4 is first removed using a chemically selective etchant, and then the shallower channel structure is etched. Since the nozzle structure area is also still open during the second etch step, it is further etched until the breakdown to the top of the silicon wafer occurs. Due to the principle of orientation-dependent etching, the etching process always produces shaped etch pits. Round nozzle openings would be better optimized from a fluidic point of view. Therefore, instead of the orientation-dependent etching, a DRIE etching process is often used for the nozzle structure.

The preparation of the second silicon substrate for the production of the passive valve and the membrane is a two-step etching process. Since the first etching step is very shallow (membrane thickness approx. 10 - 20 µm), a double mask layer system is not required. The preparation of the second silicon substrate for the production of the passive valve and the membrane is a two-step etching process. Since the first etching step is very shallow (membrane thickness approx. 10 - 20 µm), a double mask layer system is not required. After removing the etch masks, both silicon wafers are bonded together by means of silicone fusion bonding. Here, the preparation of the piezo actuator is listed only symbolically, since it is to be mounted hybrid in the technology flow. For technologies where the piezo element is to be deposited as a thin layer, it would be helpful if the second silicon substrate had a flat bottom side.

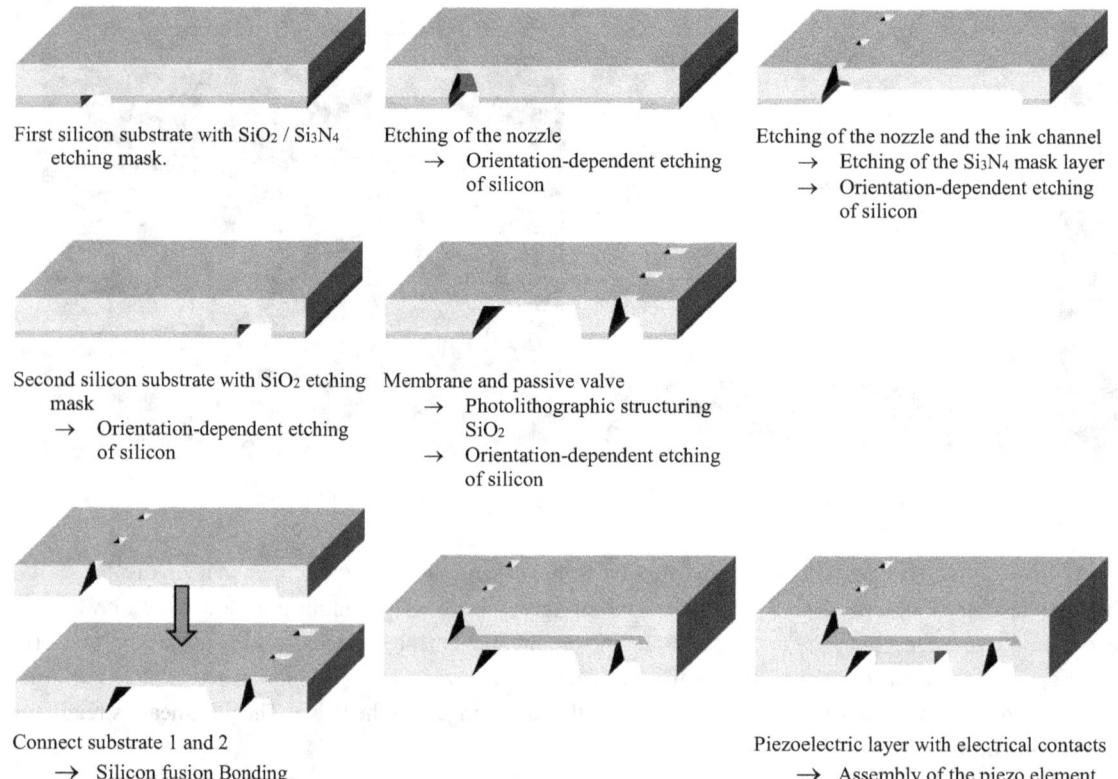

Fig. 4.24: Schematic representation of the technological steps for the realization of an inkjet head made of silicon

4.4 Acoustic MEMS

In terms of numbers, microphones are the most widely used components in acoustic MEMS. Actuators, on the other hand, have not yet managed to replace old technologies on a large scale in this area. One MEMS element that could change this is loudspeakers. They are to be used in headphones and smartphones, and replace the classic moving-coil loudspeakers. The advantage of MEMS speakers is their lower price, smaller package volume, lower mass and a full SMD compatibility. The biggest problem of MEMS loudspeakers is the radiated power. At a given frequency, it depends on the area and displacement of the loudspeaker diaphragm. Since both can only be realized to a very limited extent in MEMS elements, the first applications are loudspeakers with very low output power, as is common in earphones or hearing aids.

Classic piezo transducers are designed for higher frequencies (1 kHz - 200 MHz), and are mainly used for the generation of signal tones (piezo buzzers) and acoustic measuring waves (ultrasonic transmitters). The applications benefit from the fact that only sound waves in a very narrow frequency range have to be generated. For this purpose, the elements are excited at their resonant frequency, which allows a high amplitude of the membrane deflection, and thus a large sound power to be generated. In this case, it is also helpful in this case that as the frequency increases, the sound power also increases for the same membrane deflection. Piezoelectric transducers have a simple design, and in the simplest case consist only of a base substrate, a piezo layer and the electrical contacts. Steel or brass sheets, as well as ceramic substrates for higher frequencies (from 20 kHz), are used as the basic substrate.

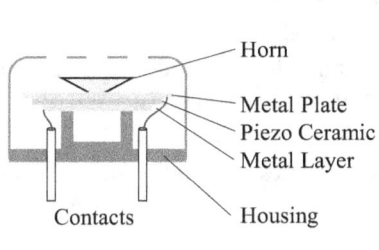 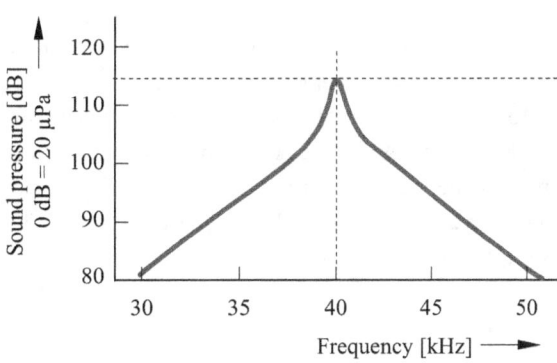

Fig. 4.25: Structure of a discrete piezoelectric ultrasonic transducer (40 kHz) and its exemplary frequency response

Since piezoelectric transducers can be used as both transmitters and receivers of acoustic signals, it is possible to use them to perform time-of-flight measurements. For this purpose, an ultrasonic signal (e.g., 40 kHz) is transmitted, which is reflected by the object to be measured. The reflected wave is detected in the receiver, whereby the distance to be measured results from the transit time of the signal.

MEMS

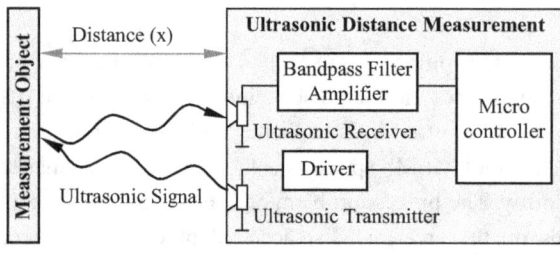 

Fig. 4.26: Structure of an ultrasonic distance meter

$$x = \frac{t_{tran}}{2} \cdot v \qquad [90]$$

with x — *Distance to the measurement object*
t_{tran} — *Measured transit time of the ultrasonic signal*
v — *Ultrasonic speed in air (342 m/s)*

This transmitter and receiver system can be used to measure distances, as well as material properties and flow velocities. The frequencies used vary from 20 kHz to 100 MHz. In principle, the accuracy of the distance measurement increases with higher frequencies, but the range decreases. Because the attenuation of acoustic waves in gases is relatively high, frequencies in the kHz range are usually used. In liquids or solids, the attenuation is much lower so that frequencies of up to 100 MHz can be used.

To save cost, space and weight, the individual components of ultrasonic rangefinders can be integrated into one device. A good example of this is the Time-of-Flight (TOF) sensor from TDK (CH101), which combines the transmit and receive functions into a single piezo element. The control and signal processing electronics are also integrated into the LGA package. The TOF sensor operates at 180 kHz and measures 3.5 x 3.5 x 1.26 mm^3.

Ultrasonic sensor arrays truly demonstrate the advantages of microsystems technology. They are not only used to measure a point distance, but also to record a 3d-image of the measuring object. A well-known example of this comprise the ultrasound devices in medical technology. Integrated transducer arrays for ultrasound imaging are mostly based on the capacitive operating principle. They are thus similar in design to MEMS microphones. In addition to detecting sound signals, they are also capable of emitting sound waves. This is achieved by cyclically applying a voltage to the membrane, which causes the membrane to vibrate and emit an acoustic wave.

Fig. 4.27: Transducer array for ultrasound imaging (Source: IPMS Fraunhofer)

Fig. 4.28: Transducer array with integrated flexible contacts (Used with permission of Koninklijke Philips N.V)

Conventional ultrasonic transmitters are not suitable for the transmission of low frequencies, since they cannot generate sufficient sound pressure at low frequencies owing to the small amount of available displacement of the membrane. When generating sound waves in the audible frequency band, softer membranes with a larger deflection are necessary. One possibility is the use of polymer membranes, such as those used by the company USound. Another way is taken by the company XMEMS. It slits the silicon diaphragm and thus obtains a larger possible deflection. The slots in the membrane must be so narrow that the lowest frequency to be transmitted does not experience an acoustic short via the slot openings. For example, the micro loudspeaker from XMEMS reaches a lower frequency of 20 Hz with a slot width of 9 µm.

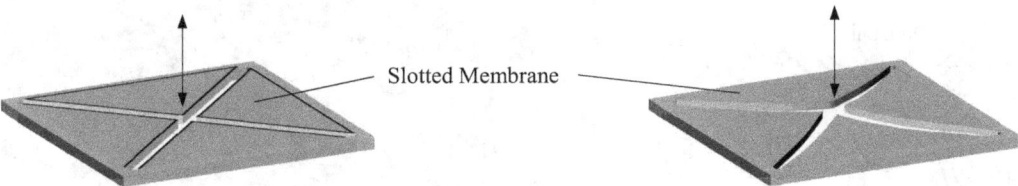

Fig. 4.29: Slotted silicon membrane of a MEMS loudspeaker

Both products presented here are relatively new on the market and are mainly used in earphones and hearing aids. At present, the micro-speakers are not yet powerful enough to replace the speakers in smartphones.

5 RF-MEMS

RF-MEMS are microelectromechanical systems that provide functions in the high frequency range. They cover frequencies from the kilohertz to the gigahertz range. Classic RF-MEMS are oscillators, RF-switches and filters.

5.1 Oscillators

The resonance frequencies of mechanical spring-mass systems are used for the realization of MEMS oscillators. Since these frequencies are relatively temperature and time stable, these serve to generate reference frequencies and system clocks in electronic systems.

The classical reference generation is done by means of quartz components. In the simplest case, they consist of a piezoelectric disk that is metallized on both sides. If a voltage pulse is applied to the piezoelectric element, the element expands (piezoelectric effect). After the voltage is switched off, the element contracts again, generating a voltage at the contact surfaces (inverse piezoelectric effect). In turn, since this generates an expansion in the piezoelectric element, the result is an electrical oscillation that corresponds to the natural frequency of the geometry of the piezoelectric element. Suitable electronic circuits can be used to continuously excite the quartz component, and thus obtain a stable output signal with a defined frequency.

The classical reference generation is done by means of quartz components. In the simplest case, they consist of a piezoelectric disk that is metallized on both sides. If a voltage pulse is applied to the piezoelectric element, the element expands (piezoelectric effect). After the voltage is switched off, the element contracts again and generates a voltage at the contact surfaces. In turn, since this generates an expansion in the piezoelectric element, the result is a mechanical and electrical oscillation that corresponds to the resonance frequency of the geometry of the piezoelectric element.

Another approach involves MEMS oscillators, in which a spring-mass system is integrated and electrically excited. Fig. 5.1 shows a MEMS oscillator from SiTime. It is structured and hermetically encapsulated in single crystal silicon by DRIE etching. The individual process steps for this are described in detail in the Packaging chapter in Fig. 2.47.

MEMS oscillators are offered in a frequency range from 8 kHz to 2.1 GHz. Their advantages over conventional quartz components are their small size, low price and an insensitivity to mechanical shocks.

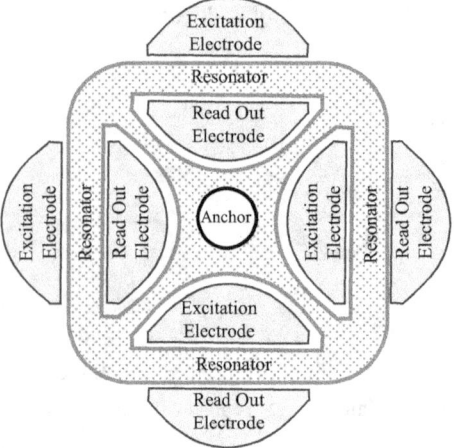
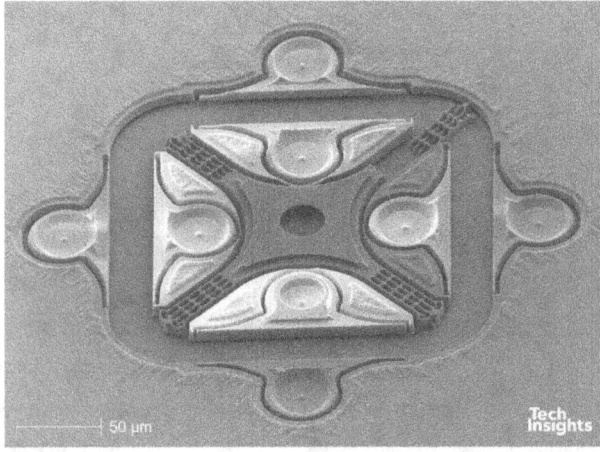

Fig. 5.1: *MEMS oscillator from SiTime (the main part of the resonator structure is missing in the right picture) (Source of the right picture: Tech Insight)*

5.2 SAW-Filters

Surface acoustic wave (SAW) filters use the propagation time effect of surface waves for filtering. For this purpose, a wave is excited on a piezoelectric substrate via a metallic finger structure, which propagates to the receiver structure. In turn, in this structure, the piezoelectric effect generates a voltage, that can be coupled out at the metal electrodes of the receiver. $LiNbO_3$, $LiTaO_3$ or SiO_2 (quartz) are used as substrate material[18].

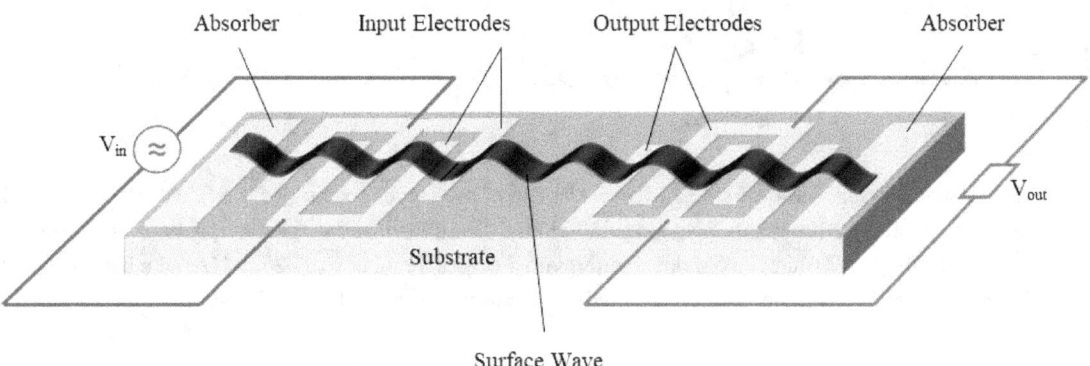

Fig. 5.2: Principle sketch of a SAW filter

In the simplest case, SAW filters have bandpass character. For this purpose, the fingers of the metal electrodes are positioned so that their spacing corresponds to half the wavelength of the center frequency of the bandpass filter. Input signals with this frequency are transmitted almost without attenuation, whereas all other frequencies are strongly attenuated due to the mismatch of finger spacing and wavelength.

$$\lambda = \frac{v}{f} \qquad [91]$$

with λ - Wavelength of the surface wave
v - Velocity of propagation of the surface wave
f - Frequency of the surface wave
$f_{bandpass}$ - Center frequency of the bandpass
Δx - Distance between the fingers of the input and output electrodes

$$f_{bandpass} = \frac{v}{2 \cdot \Delta x} \qquad [92]$$

SAW filters are excellent bandpass filters that are used in a frequency range of 35 MHz to 3 GHz. Here, the upper frequency limit is determined by the minimum finger spacing that can be produced technologically. In the lower frequency range, the structure becomes very large, so that LCR filters tend to be used here. A characteristic of SAW filters is a high ripple in the passband and stopband, which is generated by the superposition and reflection of various waves on the substrate surface.

Due to the principle, high signal powers (> 10 W) cannot be transmitted using SAW filters. In this case, LCR filters or cavity resonators are used.

---

[18] D. Morgan, Surface Acoustic wave filters, Elsevier Ltd, 2007.

MEMS

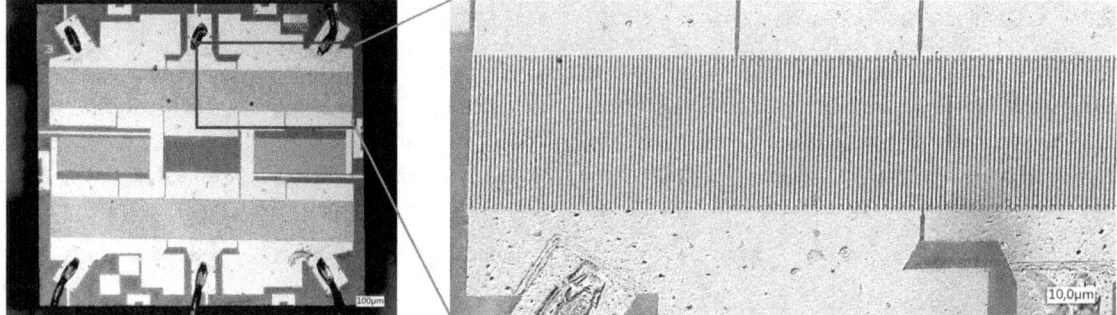

Fig. 5.3: Chip photo of a 433 MHz SAW filter

Fig. 5.3 shows the chip photo of a 433 MHz SAW filter. The finger spacing of the SAW structure is 4.5 µm, which corresponds to a propagation speed of 3900 m/s. The transmission characteristics of this filter are shown in Fig. 5.4. The SAW filter is a bandpass filter with a center frequency of 433 MHz and a bandwidth of 7 MHz. With these parameters, the filter is designed for use in receivers in the ISM band.

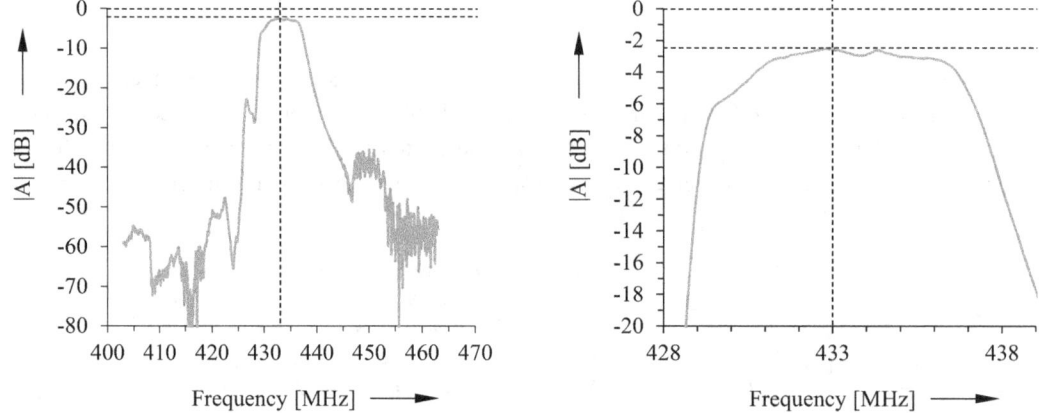

Fig. 5.4: Bandpass characteristic of a 433 MHz SAW filter [19]

SAW Sensors

SAW sensors are based on the influence of external influences on the speed of the surface waves. These can be, for example, temperature, pressure, stress, humidity or chemical sensors. The advantage of SAW sensors is that they can be read out wirelessly without their own power supply. For this purpose, a high-frequency signal is coupled in via an antenna and converted into a surface wave via the finger structures. This is reflected back at special reflectors on the SAW chip and then sent back via the finger structure and the antenna. If the excitation frequency corresponds to the distance between the finger structure and the reflectors, the returned amplitude is maximum.

---

[19] Datasheet B3710, EPCOS AG, 2015

For reading out the sensors, a transmitter emits a continuously varying frequency to determine the frequency at which a maximum response amplitude occurs. Due to the arrangement of the reflectors, it is furthermore possible to imprint a signal pattern on the signal response, which can be used to uniquely identify the response signal.

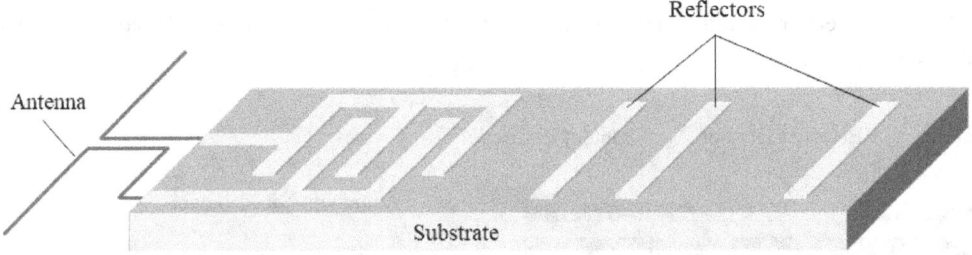

Fig. 5.5: Structure of a SAW Sensor

SAW sensors are most commonly used as temperature sensors. They are based on the variation of the resonant frequency as a function of temperature. Depending on the material used, SAW structures have different temperature coefficients. If the lowest possible temperature coefficient is sought for SAW filters, high temperature coefficients are preferred for temperature sensors.

SAW temperature sensors can be used over a wide temperature range. Commercial sensors are available from -55 °C to 400 °C. In most cases, the sensors are configured for a license-free frequency band (e.g., 433 MHz ISM band). For the 433 MHz example, the temperature coefficient of the sensors is typically 10 kHz/K[20].

The achievable temperature measurement accuracy for SAW sensors is about 0.5 °C[20]. However, temperature calibration is necessary for this, since the technologically caused frequency tolerance is 150 kHz[20], which corresponds to a temperature error of 15 °C.

Fig. 5.6 shows the application of a SAW temperature sensor as a wireless food thermometer in a household oven. In this application, the sensor is integrated into the tip of the skewer, which is inserted into the roast to measure the internal meat temperature. The temperature can be read out by radio (433 MHz) via the antenna integrated into the handle of the skewer, which thus allows the ideal control of the oven temperature.

Fig. 5.6: Wireless food thermometer from Miele based on a SAW sensor in the 433 MHz range (Source: Miele)

---

[20] Datasheet, TSE F143, SENSeOR SAS, 2013

5.3 BAW-Filters

Body acoustic wave (BAW) filters are based on the propagation of acoustic waves through a material. In contrast to SAW filters, the waves do not travel along the surface of a piezoelectric substrate, but rather into the interior of the substrate. Fig. 5.7 shows a possible BAW structure. In this structure, a piezoelectric aluminum nitride layer is used to generate and receive the head waves. The sound signals are reflected and transmitted to the receiver electrode through the underlying layer sequence with alternating layers of high and low acoustic impedance.

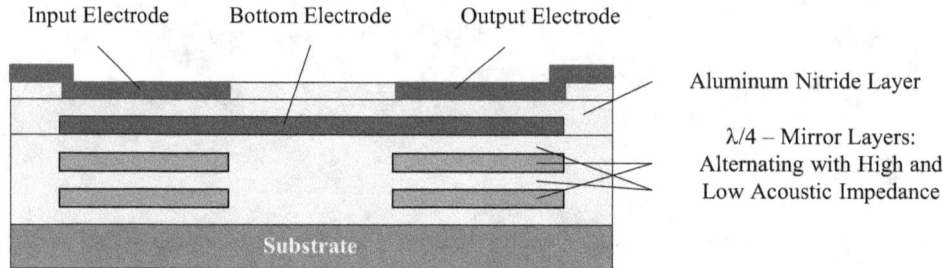

Fig. 5.7: Structure of a BAW filter

BAW filters usually have bandpass character with a center frequency of 1 GHz - 20 GHz. They are excellent as filters for WLAN and mobile radio applications. Due to their design, BAW filters are compatible with the CMOS process and can be integrated together with the active electronics on one chip. BAW filters are not practical for frequencies below 1 GHz, since layer thicknesses greater than 1 μm (λ/4) are technologically difficult to realize.

Fig. 5.8 shows a 2.44 GHz filter with a bandwidth of 79 MHz. It is designed for filtering Bluetooth and WLAN antenna signals (2.4 - 2.48 GHz) while suppressing the LTE signal (2.5 - 2.7 GHz).

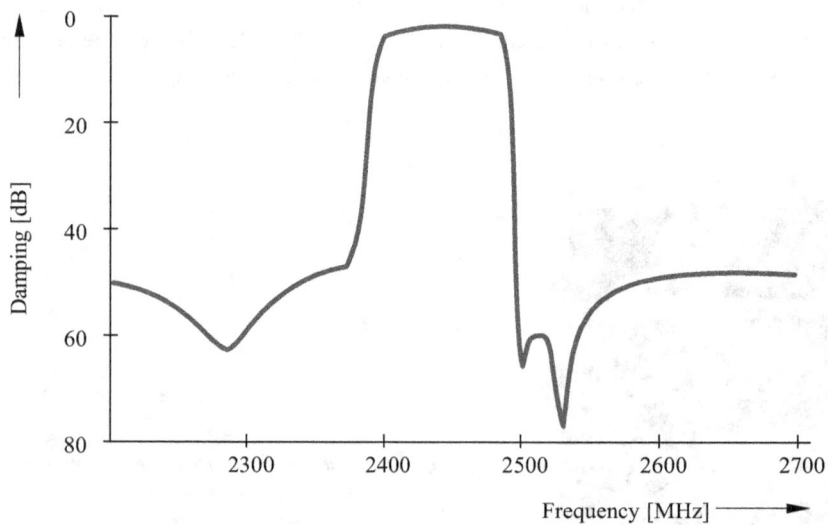

Fig. 5.8: Bandpass characteristic of a 2.44 GHz BAW filter [21]

---

[21] Datasheet B9604, EPCOS AG, 2015

6 Practical Course

Inclination angle measurements on the basis of acceleration sensors

The course should treat the use of accelerometers to determine the static angle of inclination relative to the gravity. The aim of the course is to get to know the characteristics and parameters of acceleration sensors, as well as the adaptation of a sensor for a specific measurement problem. To learn and deepen this through practical activities, a sensor board is set up, which contains a 2-axis accelerometer (ADXL202) and analog evaluation electronics. With this, both the typical parameters of an acceleration sensor and of the inclination angle measurement can be determined. The practical course is complemented by a FEM-simulation. The results of this simulation should be compared with the measured dynamic parameters of the sensor board and be evaluated.

Experimental Setup

To determine the angle of inclination, the method of measuring the effective acceleration in the x- and y-directions is used. For this purpose a sensor board is built, which consists of the acceleration sensor (ADXL202) and an analog signal-processing unit.

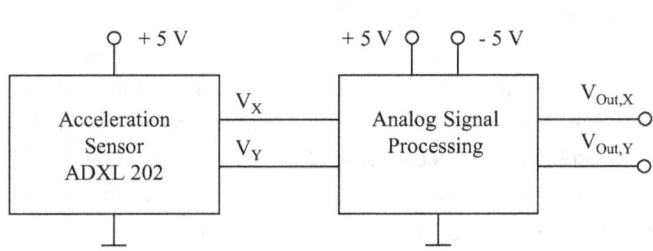 

Fig. 6.1: Block diagram of the overall sensor assembly and implementation on a breadboard

The sensor board enables a 2-dimensional acceleration measurement, and gives an acceleration proportional output voltage at the outputs $V_{Out,X}$ und $V_{Out,Y}$ with a sensitivity of 1 V/g. Here the sensor takes over the conversion of the acceleration into a voltage value and the signal-processing unit adjusts the output signal to the desired output behavior according to Eq. [93] and [94]. In the following, the output voltages of the sensor ADXL202 will be referred with V_X and V_Y, and the output voltage of the sensor board with $V_{Out,X}$ and $V_{Out,Y}$.

$$V_{Out,X} = \frac{1V}{1g} a_x \qquad [93]$$

$$V_{Out,Y} = \frac{1V}{1g} a_y \qquad [94]$$

with a_x, a_y - Acceleration in x- and y-direction
$V_{Out,X}$, $V_{Out,Y}$ - Output voltages of the sensor board

Acceleration Sensor ADXL202

The sensor ADXL202 from Analog Devices is a 2-axis acceleration sensor with a measuring range of ± 2 g. The sensor contains a surface-micromachined spring-mass system and monolithically integrated evaluation electronics. The overall sensor is integrated in an 8-pin CLCC-package.

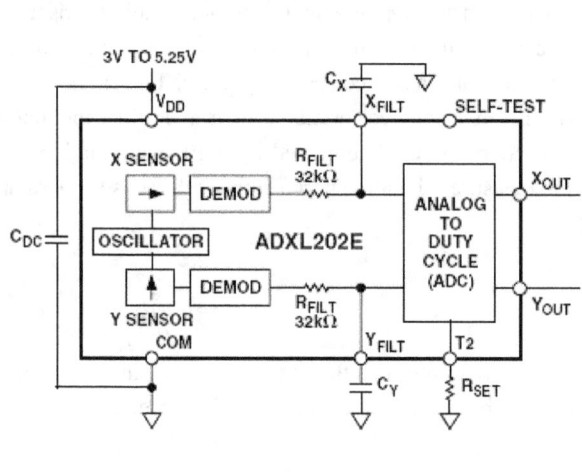

Pin-No.	Pin-Name	Description
1	ST	Self-test
2	T2	Connect R_{SET}
3	COM	Common
4	Y_{OUT}	Y-channel duty cycle output
5	X_{OUT}	X-channel duty cycle output
6	Y_{FILTER}	Y-channel, ratiometric analog output (V_Y)
7	X_{FILTER}	X-channel, ratiometric analog output (V_Y)
8	V_{DD}	Power supply

Fig. 6.2: Block diagram and pin function description of the acceleration sensor ADXL202

The sensor has two signal output options. These are a duty cycle output and a ratiometric output, each for the x- and y-axis. The two ratiometric analogue outputs are used for the realization of the sensor board in this lab work. They spend an output voltage that is proportional to the acceleration. Since they are ratiometric, the output voltage is even proportional to the supply voltage V_{DD}. In the datasheet of the ADXL202 the dependence from the output voltage is given for two commonly used operating voltages V_{DD} = 5 V and V_{DD} = 3.3V.

$V_X = V_{off,x} + S_x \cdot a_x$ [95]

$V_Y = V_{off,y} + S_y \cdot a_y$ [96]

with S_x, S_y — Sensitivity x-, y-channel [mV/g]
a_X, a_Y — Acceleration in x-, y-direction [g]
V_X, V_Y — Output voltage x-, y-channel [V]
$V_{off,x}$, $V_{off,y}$ — Offset voltage x-, y-channel [V]
V_{DD} — Supply voltage [V]

Parameter	Minimum	Typical	Maximum	Temperature Drift
V_{Off}	2.1 V	2.5 V	2.9 V	$dOff/dT = 2.0$ mg / °C
S	265 mV/g	312 mV/g	360 mV/g	$\Delta S_{max} = \pm 0.5\ \%$

Tab. 6.1: Minimum, typical and maximum parameter values for $V_{DD} = 5$ V and for the ADXL202JE

Complete Circuit of the 2-Channel Sensor Board

The sensor board consists of the acceleration sensor and a 2-channel differential amplifier. The differential amplifier compensates the offset voltage of the ADXL202 and increases the sensitivity to 1V / g.

$$V_{Out,x} = 1\,V/g \cdot a_x \quad [97] \qquad V_{Out,y} = 1\,V/g \cdot a_y \quad [98]$$

By means of Tr_1 the offset voltage can be compensated and by Tr_2 the gain of the differential amplifier (A_{Diff}) can be adjusted.

$$A_{Diff} = \left(1 + \frac{2 \cdot R}{R_3 + T_{r2}}\right) \quad [99]$$

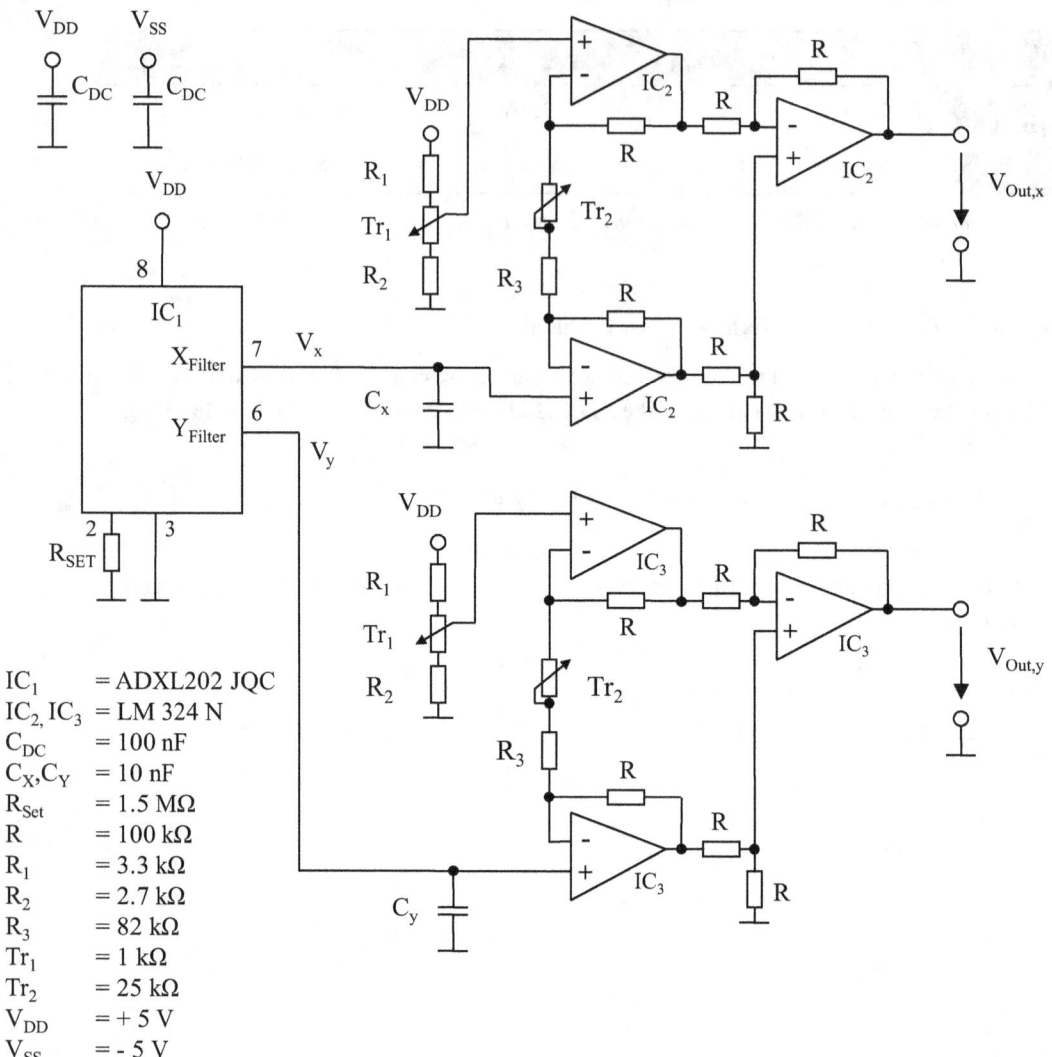

IC_1	= ADXL202 JQC
IC_2, IC_3	= LM 324 N
C_{DC}	= 100 nF
C_X, C_Y	= 10 nF
R_{Set}	= 1.5 MΩ
R	= 100 kΩ
R_1	= 3.3 kΩ
R_2	= 2.7 kΩ
R_3	= 82 kΩ
Tr_1	= 1 kΩ
Tr_2	= 25 kΩ
V_{DD}	= + 5 V
V_{SS}	= − 5 V

Fig. 6.3: Complete sensor board

To allow an easy installation of the sensor on the breadboard, the CLCC-package of the ADXL202 was soldered to a DIL adapter. The pin assignment of the DIL adapter is pin compatible to the sensor housing. Both the DIL adapter as well as the two operational amplifiers should be mounted on the breadboard using IC sockets.

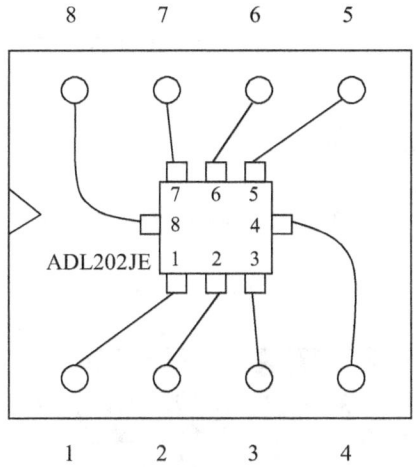 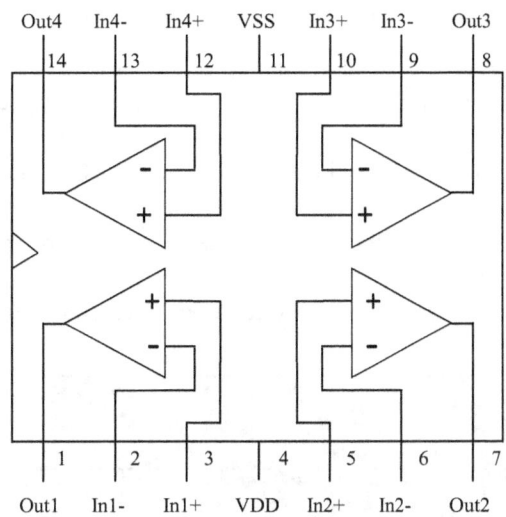

Fig. 6.4: Pin configuration of the DIL-adapter ADXL202

Fig. 6.5: Pin configuration OpAmp LM324

For the amplification, the offset correction and the buffering an instrument amplifier is used, which is executed separately for the x- and y-channel. The instrumental amplifiers are realized with the LM324, which provides four OpAmp in one IC package.

In addition, the LM324 has the following properties:
- Internally frequency compensated
- High differential gain (100 dB)
- Large supply voltage range (3 V to 32 V)
- Low power consumption (700 µA)
- Low offset voltage (2 mV)
- Low input current (45 nA)

The circuit is completed by two capacitors (C_{DC}), which buffer the voltage against transient current peaks. On blocking capacitors for the individual ICs has been omitted in this approach.

Startup Procedure for the Sensor Board

The assembly and the startup of the sensor boards should be done stepwise in order to identify easily possible errors. The following procedure is recommended:

1. Soldering the circuit board without the OpAmp and ADXL-accelerometer
2. Test of the current consumption
3. Attaching the acceleration sensor, test the power consumption, test the angular sensitivity on V_X and V_Y
4. Attach the OpAmp, test the power consumption, test the angular sensitivity on $V_{Out,X}$ and $V_{Out,Y}$

	Without ADXL Without OpAmp	With ADXL Without OpAmp	With ADXL With OpAmp
Current consumption at + 5 V	1.4 mA	2 mA	3.5 mA
Current consumption at - 5 V	0 mA	0 mA	1.4 mA

Tab. 6.2: Typical values for the power consumption of the circuit in different configuration variants

Offset adjustment

For offset calibration, the sensor board is brought into horizontal position (rotary table Fig. 6.6) and then compensated by trimmer Tr1 for both channels. This gives $V_{Out,\,x} = V_{Out,\,y} = 0$ V.

Adjustment of the Sensitivity

For this purpose, the sensor is placed in the vertical position and each output is adjusted to 1 V by means of Tr2. By rotating the sensor to 180 degrees, an acceleration of -1 g occurs and theoretically, a voltage of -1 V is now observed at the output. But a slightly different value will be measured. The reason for this is the erroneous tilting of the sensor chip against the board level. The calibration should be carried out so that the deviations from the ideal value of the output voltages (± 1 V) are equal in magnitude in the two endpoint of the characteristics

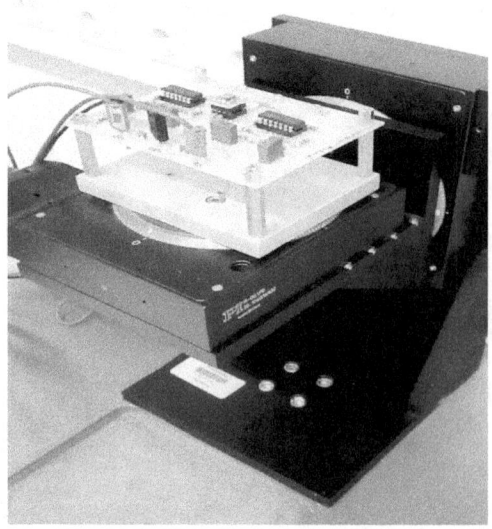

Fig. 6.6: 2-axis precision rotary table for calibrating the zero point and the sensitivity as well as for measuring the angular dependence

Measurements of Static Parameters

Sensitivity of the Acceleration Sensor

The sensitivity of the sensor system is ideally 1 V/g after the calibration. The sensitivity is measured in the vertical position at which ± 1 g acts on the acceleration sensor. To compare the calculated sensitivity with the sensor data sheet, the adjusted gain (V_{Diff}) of the instrumentation amplifier must be determined. The sensor sensitivity (S) is obtained from:

$$S = S_{gesamt}/A_{Diff} \qquad [100]$$

The gain of the instrumentation amplifier (A_{Diff}) is determined in two different sensor positions. For this, the output voltage of the sensor board (V_{Out}) and of the sensor ($V_{x,y}$) has to be measured.

$$A_{Diff} = \frac{\Delta V_{Out}}{\Delta V_{x,y}} \qquad [101]$$

The determination of the gain can only be done after the calibration of sensitivity and is carried out separately for each channel.

Measuring the of the Alignment Error

Mounting the sensor in the package results in a misalignment of the sensor coordinate system with respect to the outer edges of the package. The misalignment can be described by three angles (γ_x, γ_y, γ_z). In the data sheet of the ADXL202 only one value for the alignment error is given with ± 1 degree. Furthermore there can be a significant increase of the misalignment occurs by soldering the sensor onto the test board.

For the measurement of the alignment error (γ_x, γ_y) the zero point value is measured in the 0° position as well as the 180° position.

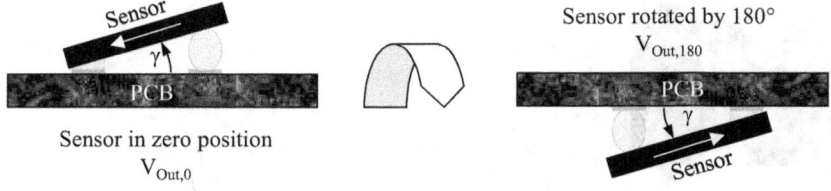

Fig. 6.7: Measurement procedure for determining the alignment error

$$\gamma_x = \arcsin\left(\frac{1}{2} \cdot \frac{V_{Out,x,180°} - V_{Out,x,0°}}{S_a \cdot 1g}\right) \qquad [102] \qquad \gamma_y = \arcsin\left(\frac{1}{2} \cdot \frac{V_{Out,y,180°} - V_{Out,y,0°}}{S_a \cdot 1g}\right) \qquad [103]$$

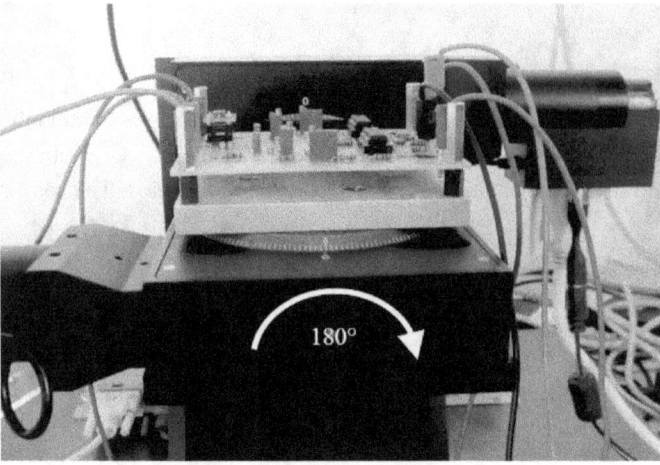

Fig. 6.8: Determining the alignment error (γ_x, γ_y)

To determine the alignment error (γ_z), the rotary table is brought into a vertical position and the x-axis into a horizontal position. The rotary table is then rotated 180° around the z-axis.

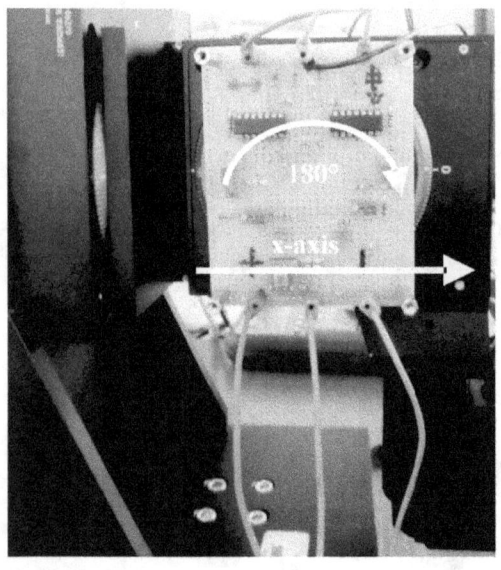

Fig. 6.9: Determining the alignment error (γ_z)

$$\gamma_z = \arcsin\left(\frac{1}{2} \cdot \frac{V_{Out,x,180°} - U_{Out,x,0°}}{S_a \cdot 1g}\right) \qquad [104]$$

Determination of Dynamic Parameters

Dynamic measurements show the behavior of the sensor as a function of the frequency. Capacitive acceleration sensors based on spring-mass systems have a low-pass behavior of the second order. For the acceleration sensor ADXL202, the producer gives a resonant frequency of $f_R = 10$ kHz. However, a far more dominant influence on the frequency response of the sensor board has the vibration behavior of the printed circuit board and the electronic low-pass behavior of the evaluation electronics. Fig. 6.10 represents the frequency response of a sensor board and the associated vibration modes of the PCB.

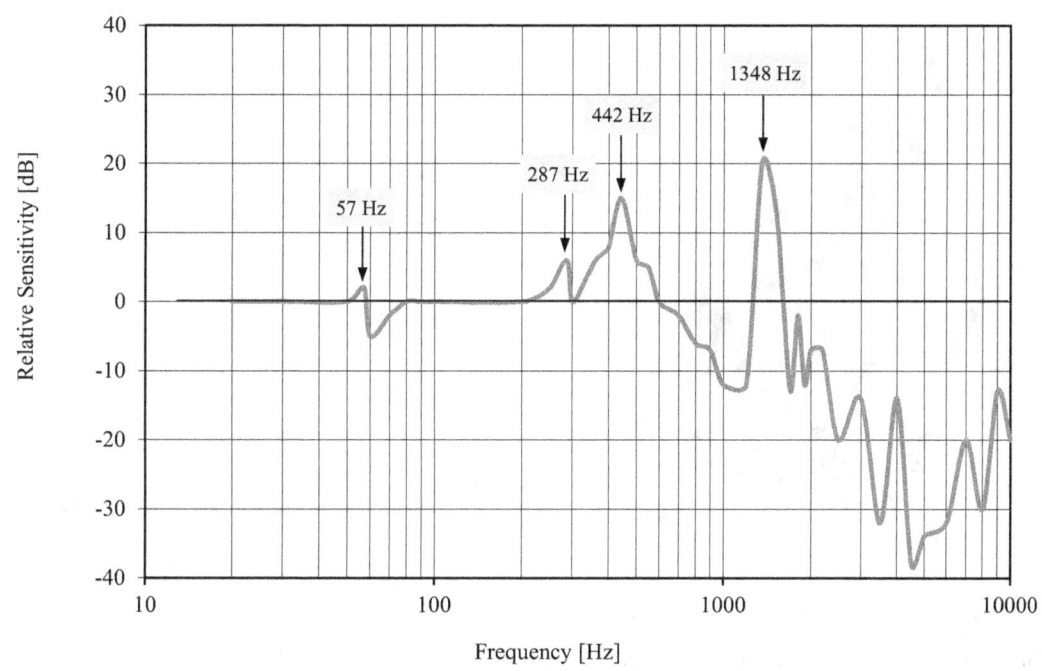

1. Eigenmode $f = 60$ Hz

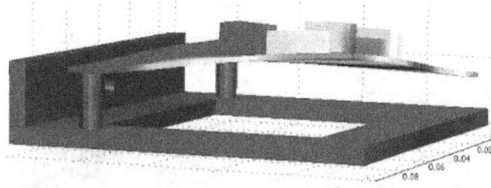

2. Eigenmode $f = 250$ Hz

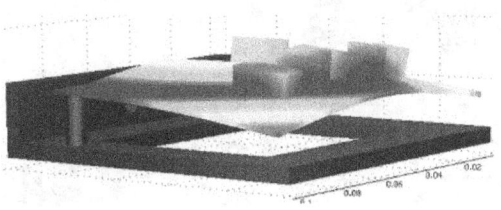

4. Eigenmode $f = 487$ Hz

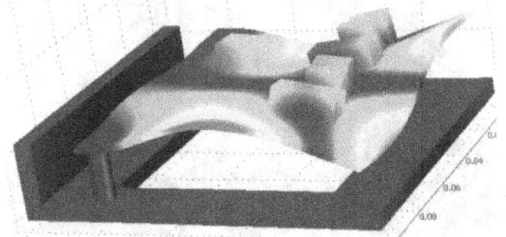

9. Eigenmode $f = 1270$ Hz

Fig. 6.10: Frequency response and associated eigenmode (FEM-simulation) of a mounted sensor board

For the measurement of the frequency response, the sensor board is excited by means of a vibrating table (shaker) and the measured output data are compared with a reference acceleration sensor. The sensor board have to be fixed for this measurement at only two points, in order to achieve the highest possible impact of PCB-resonances. This very loose suspension would be rather unsuitable for acceleration measurements in practice, but it gives a better assignability of the simulated eigenmodes to the measured resonance points. The reference sensor is a piezoelectric acceleration sensor having a sensitivity of 1 V/g in the selected setting of the charge amplifier.

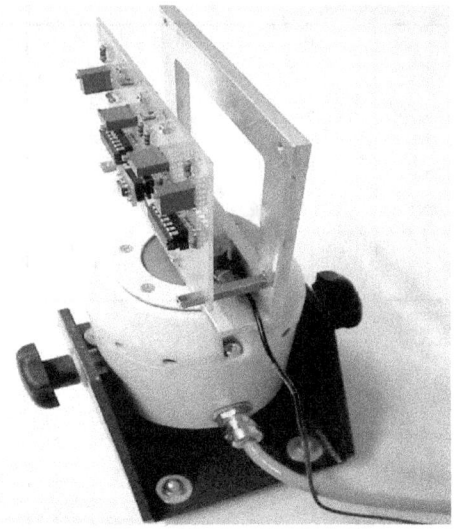

Fig. 6.11: Assembly of the sensor board on the shaker

Measurement of the Temperature Dependence

The temperature dependence of the offset voltage and the sensitivity should be determined only for one channel of the sensor. For this purpose, the climate chamber is successively set to different temperatures (see Quick Start Guide on climate chamber) and after reaching the respective temperature and a waiting time of 5 minutes the output voltage values in the vertical and horizontal position of the sensor are measured.

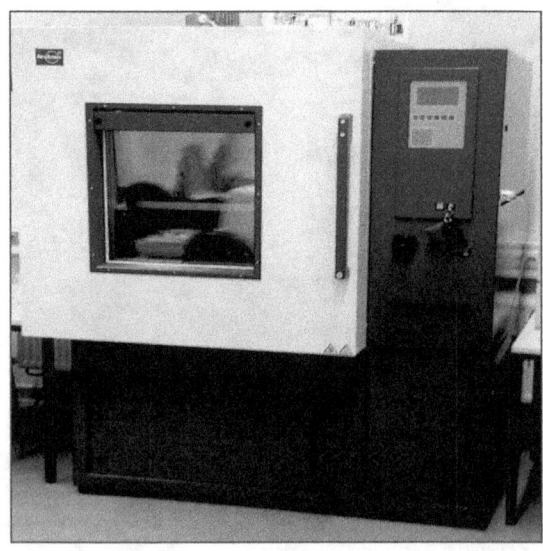
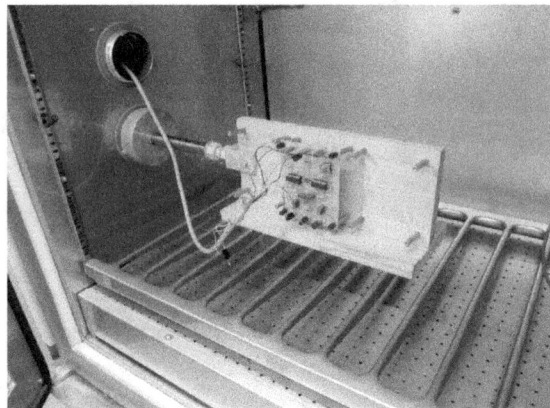

Fig. 6.12: Fixture for receiving the temperature dependence of the sensor offset voltage and of the sensitivity

During the recording of temperature dependence, there can be arise a hysteresis between the heating and the cooling curve. The reason for this is that with a latency of only 5 minutes, the oven temperature and the temperature of the sensor still differ. For the determination of the temperature parameter, therefore, the respective mean values of the heating and cooling curve are used.

The temperature coefficient of the offset [mg/°C] of the sensor element can be determined from:

$$\alpha_{Off} = \frac{1}{S_{25°C}} \frac{\Delta V_{Off}}{\Delta T} \qquad [105]$$

For the maximum percentage change in sensitivity (ΔS), an S is determined for each temperature, and then the minimum (S_{MIN}) and maximum value (S_{MAX}) of the sensitivity is calculated. The difference between these two values set in relation to the initial sensitivity at 25 °C ($S_{25°C}$) gives the parameters (ΔS).

$$S_{MAX} = \frac{\left|V_{Out,vertical}(T) - V_{Out,horizontal}(T)\right|_{max}}{1g} \qquad [106]$$

$$S_{MIN} = \frac{\left|V_{Out,vertical}(T) - V_{Out,horizontal}(T)\right|_{min}}{1g} \qquad [107]$$

$$\Delta S = \frac{S_{MAX} - S_{MIN}}{S_{MAX} + S_{MIN}} \cdot 100\% \qquad [108]$$

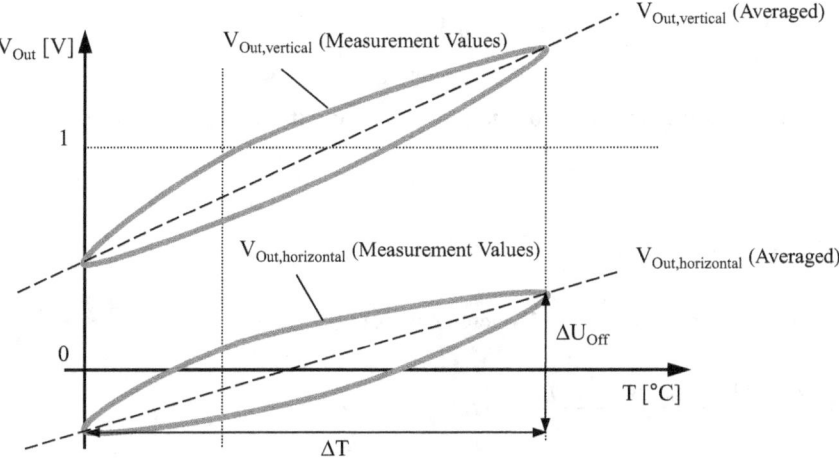

Fig. 6.13: Temperature dependence of the output voltage in the horizontal and vertical position

Inclination Measurement

For the determination of the error in the measurement of the inclination angle, a curve from -90° to +90° should be recorded in 15° increments.

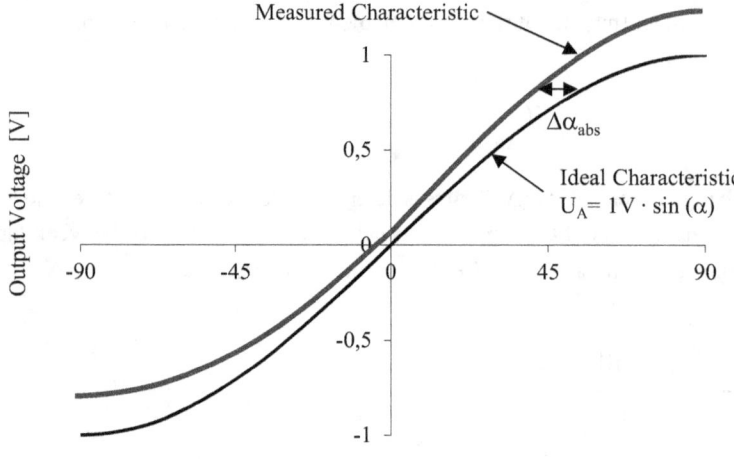

Fig. 6.14: Definition of the absolute measurement error

Ones determines the absolute measurement error $\Delta\alpha_{abs}$ as the difference of the measured value and the ideal characteristic curve.

$$\Delta\alpha_{abs}(\alpha) = \left| \alpha - \arcsin\left(\frac{V_{Out}(\alpha)}{1V}\right) \right| \qquad [109]$$

The angular sensitivity (S_α) describes the sensitivity of the sensor output with respect to the change of the angle. Ideally, it is:

$$S_\alpha = \frac{dV_{Out}}{d\alpha} = 1V \cdot \frac{\pi}{180°} \cos(\alpha) \qquad [110]$$

For the metrological determination a linearization in 15°-increments is to be made.

$$S_\alpha(\alpha - 7{,}5°) = \frac{\Delta V_{Out}}{\Delta\alpha} = \frac{V_{Out}(\alpha) - V_{Out}(\alpha - 15°)}{15°} \qquad [111]$$

Task

Startup of the Sensor Board

a) Assemble the circuit of Fig. 6.3 and proceed through it step by step:
   1. Assemble the circuit without the OpAmps and the ADXL-sensor
   2. Test the current consumption
   3. Insert the acceleration sensor, test the current consumption, test the angle sensitivity
   4. Insert the OpAmps, test of the current consumption, test the angle sensitivity

b) Set the output voltages of the differential amplifier in the horizontal position roughly to 0 V.

c) Set the output voltages of the differential amplifier in the vertical position roughly to ± 1 V.

Measuring of the Static Parameters of the Acceleration Sensor

d) Install the sensor on the turntable. Compensate the zero point offset and adjust the sensitivity (S) for both channels to 1 V/g. Document the values of the supply voltages used for the measurement.

   Note: Before measuring, check the zero position of the PCB using a spirit level in the x and y directions. Recalibrate the measuring table if necessary.

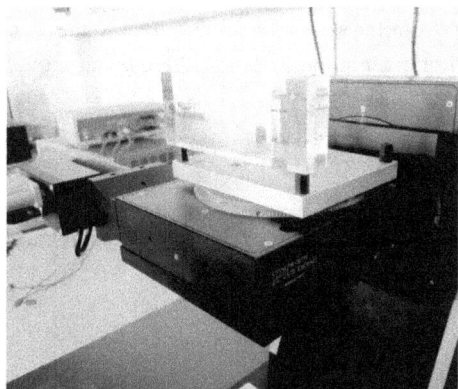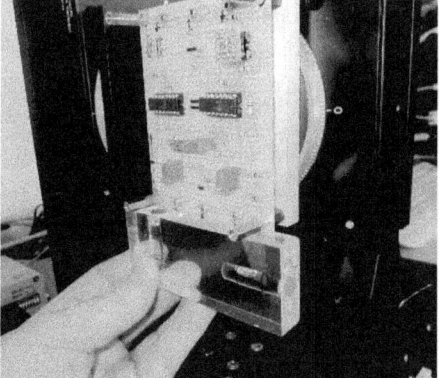

Fig. 6.15: Measurement of the zero position in x- and y-direction

   Note: Calibrate the zero point to $V_{Out} = 0\ V$ and $\pm 90°$ to $\pm 1\ V$ with symmetric deviations from the ideal value

   Note: After the calibration, the measurements of inclination dependencies should be subsequently carried out.

e) Determine the gain of the two differential amplifiers.

f) Determine the sensitivity of the sensor element (S) in the x- and y-direction and compare it to the data sheet specifications.

g) Determine the alignment error ($\gamma_x, \gamma_y, \gamma_z$) of the sensor element to the board level and compare it to the data sheet specifications.

Measuring of the Dynamic Parameters of the Acceleration Sensor

h) Mount the sensor board on the shaker (Fig. 6.11) and document the installation position of the sensor by a photo.

   Note: Tighten the screw connections (hand-tight, without tools).
   Note: Make sure that the same position is used as in the FEM simulation.

i) Measure the output voltage of the x-channel (excitation direction) in the frequency range from 10 Hz to 10 kHz at an acceleration of 0.1 g (amplitude). Choose automatically appropriate frequency steps.

   Note: Use the function „Ratio 1→2" of the oscilloscope.
   Note: The excitation voltage can be increased for higher frequencies to measure large negative attenuation values.
   Note: Measure both the minima and maxima of the transfer function as well as the values at which the attenuation returns to the initial values before and after the extremes.

j) Plot the sensitivity as a function of the frequency. Scale both axis logarithmically.

k) Compare the determined sensitivities with the static sensitivity specified in the previous section.

l) Determine the eigenfrequencies of the overall system and comparing it to the eigenmodes obtained from the FEM simulation. Represent four eigenmodes and note there eigenfrequencies in the measuring curve of the frequency dependence.

m) Draw additionally the theoretically expected frequency response into the diagram.

n) Measure the noise voltage of the X-channel and represent it graphically (screenshot form the oscilloscope). Determine the effective noise voltage and compare this with the information in the data sheet.

   Note: Use a 1:1 probe and the "High-Resolution" setting of the oscilloscope for the measurement
   Note: Use RMS-function of oscilloscope to determine noise voltage
   Note: To avoid measurement errors due to static offset values of the sensor, use the AC mode of the oscilloscope for the noise measurements

Measuring of the Thermal Parameters of the Acceleration Sensor

o) Assemble the acceleration sensor in the climatic chamber.

p) Measure the output voltage of the channel in the direction of tilt-sensitivity in vertical and horizontal position at temperatures of T = 25, 0, 25, 50, 75, 50, 25 °C. Plot both temperature curves.

   Note: Plot both temperature curves (horizontal and vertical) on the same diagram.

q) Calculate the temperature coefficient for the offset voltage and the maximum change of the sensitivity.

r) Compare the determined values with the data sheet specifications.

Application of Accelerometers for Tilt Measurements

s) Measure the dependence of $V_{Out} = f(\alpha)$ for both channels. Determine the dependence in 15°-steps.

t) Plot the dependence of $V_{Out} = f(\alpha)$ and $S_\alpha = f(\alpha)$. Additionally draw the ideal sensor characteristic into both diagrams.

u) Calculate the absolute error of the inclination measurement for the different inclination angles.

v) Mount the sensor on the inclined plane as shown in Fig. 6.16 and determine from the output voltages of the two channels (V_x, V_y), the tilt of the plane against the horizontal. Perform the measurement twice, each time with the angle rotated by 180°, in order to eliminate the influence of the misalignment of the tabletop. Calculate the average value from both measurements.

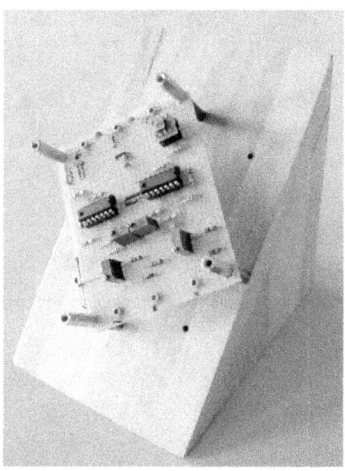 

Fig. 6.16: Installation position of the sensor board to measure the tilt of the plane and 180°-rotated position for the 2nd measurement

Masthead
Author and Publisher: Prof. Dr. Dirk Zielke

Contact Address: University of Applied Sciences Bielefeld
Interaction 1
33602 Bielefeld
Germany

email: dirk.zielke@fh-bielefeld.de

www.ingramcontent.com/pod-product-compliance
Lightning Source LLC
Chambersburg PA
CBHW080457220526
45465CB00006B/2301